HEART NOTES

DIANNA AMMONS JOHNSON

THE PRIMAVERA PRESS

ISBN: 978-1-7329991-2-1

In Appreciation

To my son, Spencer Case Redburn, who paid me a heartfelt compliment by urging me to compile and share my writings

To my husband, Leonard Johnson, who 1) suggested the title for the volume, "Heart Notes," 2) carved a heart on an ancient bone we found while hiking in Unalaska, the photo of which provides a fitting cover for the volume, and 3) (bless his heart) provided the initial editing to correct, among other things, my dyslexic spelling errors

To Grace Sheridan, who provided pivotal inspiration, through sharing of her own remarkable poetic writings and what is a joint love of everything Down East

To Gerard NeCastro, who graciously provided the path toward publication and made it all happen.

Table of Contents

Love of Nature

LOVE OF NATURE

PLACE OF THE HEART

There is a sanctuary
Embedded deep within each of us
A place known only by its keeper

The single soul residing within its embrace
Is thereby defined, nurtured, healed, and inspired.

Separate as all human beings are,
But connected by sensual ties
To all other living beings.

It is by these tethers that our hearts and minds
Find a balance and true meaning.

POETRY, PERHAPS

Maybe our first words were primitive poetry
Grunts and growls, barks, sighs, whimpers, and coos
Emotions and intentions expressed unadorned
Without syntax or clutter, but abundantly rich with meaning.

Straight-to-the-heart
Notions and messages
Binding us to one another
Through sounds and quiet spaces between sparse words
Specific to our species

Now we are overwhelmed by the cacophony of multitudinous words
Discordant clanging of a thousand bells with little meaning
Even through this deluge it is possible to reach back to the calm of
 our poetic core
To seek the simple enlightening phrases and powerful pauses
That sing to our heart

ATTEND

Words spoken by the wind
Signs painted in shadowy blues, misty greys across the sky
Shapes along the horizon, spelling out the interplay of lands,
Those within your purview and those beyond

Small sounds, hidden among the reverberations and distractions of
 larger movements
Sweet breezes stroking your cheek, almost lovingly.
Receive them all within your embrace
As you learn of the world.

THE SKY

The sky with its boundless shroud of blue
And fleecy white puffs of cotton
Stands alone, never ending
Almost forgotten

In Sky Lake

Lord have mercy! It's Sky Lake in the morning. Tiny little nothing of a lake, one campsite beat to death, and bugs, bugs, bugs, swarming in your face, in your ears, in your mouth, and up your nose. Even so, I would not choose to be elsewhere, partly because every other human in this world *has* chosen to be elsewhere, allowing me to enjoy the little lake's unruly charm in solitude. It sits puddled against a sheer, gray mountain peak studied with patches of snow, a rather pretentious backdrop for such a little lake. My only companions are the beautiful cutthroat trout that swim in its waters and the moose that joined them early this morning for a paddle across the lake.

I have devoted much of the day to studying the lake, concentrating on the miraculous changes that are played out on its surface, which is sometimes smooth and plain, a perfect reflector for the grand scheme around the shore; other times, covered in choppy waves. It seems like a single organism with a life and soul of its own, an animal soul because its essence is movement. It is capable of a single coordinated response to some things such as a strong wind that can beat up waves like icing on a cake. Individual waves act like a living extension of the main body, extensions that appear to have gone mad, pulling themselves up from the lake. Each wave is a water leap, throwing itself, flinging itself in a curved hook to catch hold of the wind.

When the breeze dies down into wispy and intermittent puffs, the light shows a more complex character. Certain areas become smooth and still with a thin line of ruffled water, swirling across like a curving ribbon. All of these shapes and forms are, of course, just a pleasant deception. The water itself is clear, so very clear, that looking straight down into its depths I can see details of each submerged log and stone without distortion. But as I look out over the water in an angled view, mysterious images are formed as the interface selects and rejects favored wavelengths in a playful game with the light. And it is always a dazzling show. The uneven surfaces of the water are formed and dissipated so quickly that they give only a momentary flash of light, a single sparkle of sunlight, added to countless other sparkles so that the surface becomes transparent and shining at the same time.

As an escape from the mosquitoes and relief from the sweat and grime that have accumulated after several days of hiking, I slip out of my clothes and lower myself off a log and into the lake, puncturing its shimmering surface. In one quivering instant I am swallowed whole and pulled into the body, the belly of the lake. Its coldness grips me, surrounds me, squeezes me with such a shock that my lungs jam. They are frozen and will not expand. They generate only enough breath for a few yelps. With having never let go of the log, I climb out of the water back to the safety of warm air. I slowly regain the gentle rhythm of normal breathing and am given over to a most delicious sensation. The dramatic shift from lung-wrenching chill to the warm pleasure of the afternoon sun, produces a whole-body tingle. Hmmm – mellow.

A swirling patch of rippled water forms a question mark on the lake: "Want to try it again?" I contemplate the proposition for a while in order to appreciate its full meaning. It is after all an important question. It asks about courage. This ignored bit of Montana wilderness offers a challenge, a chance to experience an initiation that is cold, harsh, and cleansing. It is an initiation into a deeper knowledge of who I am. The message is sent as an icy shock wave through skin and muscle straight through to the visceral soul, and the message is: you are truly alive. The quick broad sweep from pain to pleasure jolts the body and spirit, exposing the full extreme of human possibilities which are written in waves and wispy messages on the lake: the playfulness of sun-sparkled waters, the sensual flow of currents dancing in arabesque, the brave leap of white-capped waves, the quiet pools that reach deep and true. With these promises before me, I remain poised as the lake waits in silence for the sound of the second splash.

ELK AWAKENING

There is a sweet slumber that can surround me
Womb-like, dark and insulating
Not separating me from the world
But taking me deep within.

Usually, it finds me after solitary meandering days
Through backwoods trails, along snow-melt streams or alpine cirques.
I string up my hammock at end of the day
And with the darkness, rest comes as a treasure.

During one such evening near a jewel blue lake in Idaho
The dead of night was suddenly split open, jolting – jarring me awake
An eruption of wild movement, large animals crashing through the woods around me
Bumping against my hammock as if it were an island in the streaming crush of the herd.

The next instant they were gone, leaving a wake of deafening silence.
"Elk." I eventually managed to gather the thought out of my startled panic.
Perhaps they were my dream – but no.
Their intense muskiness still lingered; their short grunts echoed in my head.

This made me smile.

This glimpse of the secretive business of night creatures
Etched deeply in my memory by the sharp and sudden boundary of my awareness.
A gift, a reminder of the ceaseless enterprises of wilderness lost in our somnolence.

SEA MEDICINE

There is a rhythm to the roaring. The low rumble from the ferry engine fills my ears, but I can also hear a higher pitched, mechanical hum – almost an octave above – which gets louder then fades, creating an ebb and flow of sound that is almost a melody. The sound vibrates deep in my chest touching, it seems, the very center point of my feelings. I have chosen to come to the back of the ferry because it offers solitude and a lingering view of San Juan Island. Strangely, the island looks unchanged from when I approached it this morning, but of course it is I who has changed.

The day began, in a light drizzling overcast, what they like to call "rain" around Seattle. The cove near the ferry dock was deserted by other humans. It was left to me with only some tiny shore birds for company. The gulls were tied up in a conversation at the pilings near the dock – a bird atop each piling; a piling for each bird – and they were busy gawking and gossiping. A lovely crane flew over overhead, circling me twice, and announcing my arrival with its shrill screech. The beach was rock-strewn with pebbles polished like gemstones by the sea. Bits of pottery and glass – who knows from where – were mingled among the stones. They too had been rubbed and refined by tumbling in salty waves. Pieces of clear glass look frosted after this treatment and they lay about like pieces of ice chipped from a frozen pond. One such chip caught my eye. It was a good-sized fragment about two inches across, roughly square in shape, and slightly curved. There were raised letters on its convex surface and the style of the lettering suggest a manufacturing date sometime in the last century. It read:

"Dr. M
estor
ervous"

Although the words were incomplete, the message seemed clear:

"Dr. M (Dr. Moriarty's?)
Restorative tonic
Cures any nervous condition"

A quaint message from the past that potions can restore one's mental health. Nevertheless, it is the same notion upon which modern pharmaceutical houses have been founded and have flourished. It is likewise the basis for my own approach to research of the brain and its inner workings. It presupposes that the brain can be modified or restored (in the case of the tonic) by exposure to appropriate chemical concoctions. The most effective substances turn out to be chemicals which are normally present in brain, but which may get out of balance for one reason or another. The emphasis is on understanding the brain's makeup and then using normal constituents (or their substitutes) to help normalize the brain's activities. The approach is one of an intrinsic nature: help the brain heal itself.

As I studied my little gift of glass from the sea, it occurred to me that there was also another quite different approach for treating maladies of the mind: the extrinsic approach. The powerful workings of the ocean had taken this little broken shard, and carried it on swift currents, and eddies, smoothing it sharp edges and restoring it to a lovely little chip complete in itself again. Even the abbreviated message was still discernible. A discarded fragment had been turned into a small treasure by giving itself to the sea and riding the waves to new destinies. This change had been accomplished through an external power, in many ways distinctly unlike the nature of the glass itself. The restoration relied on forces beyond, outside the glass itself. Perhaps the same notion might apply to living objects, as well to humans in particular. The notion was foreign to me, as foreign as the ocean forces themselves. Although the water of the sea is the basis for our own life's juices, our salty blood, and birthing fluid, the evolutionary link has been stretched over so many eons that it no longer seems familiar. I did not feel that I was part of the sea or that it was part of me. Rather its composition and the force it generates on the open waters or along the shores, seemed strange and exotic. And I wondered if something as foreign is this had the power to change my life, to reach my soul and etch out new meanings.

The last of the ferry horn startled me as it called out to its passengers, giving them a chance to ride the sea. And as the last mooring was set free, the healing began. I had a gentler ride than my little piece of glass, but I felt the same soothing power, sharp-edged feelings were buffeted and worn down, and the frosted glass became the lens through which I viewed the world as through a sea spray mist.

With this new vision I first saw the jeweled San Juan Island, it too buffeted and sculpted by the ocean. It welcomed me in as a friend to experience a day of intense pleasures – to see the cloud curtain part and bring forth a brilliant sun-filled day; to smell the sweet green island forest; to share the wooded trail with a beautiful doe and her two fawns; to delight in the bald eagles that soar above rock bluffs. But the greatest gift was to hear the breathing of orca whales as they trolled the coast for salmon. They came close to shore at Lime Kiln Point; three bulls leading the way, followed by the females and a pair of young, that surfaced side-by-side like skipping children holding hands.

Treasures, treasures, each one, surrounded and nourished by the sea. The rhythm and balance of nature is not subtle here; it is a real and undeniable part of the Island's life. And through the lens of my glass treasure, I look to the sea to become part of my life, to erode the many meaningless layers and establish a natural balance there. The rhythmic roar of the ferry engines sound melancholy as the tree-lined ridges of San Juan Island slip from me into a distant blue haze. I am pulled away while the Island remains anchored in the depths of Puget Sound. Maybe it's just waiting. I want it to be waiting for my return.

GRAND VISTAS

One more climb of the rugged terrain which forms the south wall of Aspen Valley at the confluence of the Snowmass and Roaring Fork Rivers. I seek one last glimpse, a longing look at the horizon-to-horizon glory of the mountains. Having found an easier ascent today, I reach breathtaking views with much less effort than the previous day's trudge through scrub oak and brush to reach the summit. I have followed the crisscross of human and animal trails, littered with droppings of chipmunks, ground squirrels, marmots, and the like, and also deer tracks that puncture the soft red dirt. The bareness of the lower mountains gives way to a ragged stand of pine cedar and fir; stunted twisted trunks desperately clinging to rock ledges. In the quietness of these desolate-looking trees, a coyote crossed my path, twenty feet or so ahead of me. She moved at a cautious but not hurried pace, paused for the briefest moment on the trail and glanced my way. It is always a shock to meet such a beautiful and wild creature so unexpectedly. In a riveting instant it can send a sudden bone-chilling shiver to the soul and suck the breath away, leaving one frozen, suspended, momentarily mindless. It can, but in this case, it did not. What formed in my mind was not fear but questions. Have I intruded? Have I disturbed her early morning hunt? Have I flushed her from her den? And a sadness follows from momentarily feeling like an outsider on these mountains. The coyote is an important animal totem, and I wait through this suspended instant in time for the message she might bring to me. The wily coyote, the trickster. Is there mischief on the trail ahead from others or from my own foolish actions? The instant passes, the coyote continues confidently without breaking stride, without scampering away. I have not frightened her nor she, me.

Acceptance. It is a small but welcome blessing. "We are here together," the coyote says to me. "Let us continue on the paths we have chosen."

But I stopped for a while, just to feel the sweetness of the moment, to sit quietly and gather my goodbyes together before dispersing them down the valleys. Slowly, the silence I felt so strongly fades as rocks and animal trails begin to come to life around me with small creatures, cautiously at first and then with true abandonment, scurrying, chirping, and generally making a fuss, a comical show. Lots of darting to-and-fro, rock-scampering, tail-twitching. As I glance up from the furry creatures cavorting at my feet, my eyes begin to change their focus to a far vision, drawn again to the grand vistas, to the unbelievable softness that distance bestows on the mountains. The tawny suede look of barren rock, the deep green velvet of fur and spruce forests. Even the clawed scars of ski slopes appear as light, yellow green ribbons draped over the mountain tops. The course of rivers and creeks can be followed by the presence of river birch, which cluster there and form a playful scribble on the landscape. It all looks so soft. As if I could fling myself out and bounce on the spongy-looking mountain peaks and slide along the suede and velvet textures that seem so real in the farsighted view.

The distance distorts, I know, but the loss of detail of sharp edges, of rigid form, allows what is impossible under close inspection. The far view, the appreciation of the grand scheme of things. It is a softened view that permits me to take in broad panoramas, an appreciation of the grand plan of nature. From the sweep of the mountain range and downward plunge to the valley, I can find my way, plot my course, find my direction. From these soft outlines of mother earth's shifting tectonic plates, there is a gentle force that opens my eyes to my place in the world and lifts me up to fill it.

LOOKING FOR BEARS

A beautiful day is set out before me – gloriously free of any schedule – just me and a mountain and, according to local café gossip, a few bears. Hiking along the winding blacktop road would be pleasant enough. A small stream follows the roadside for much of the way. But there are some interesting signs – subtle, half-hidden signs – of a Jeep trail branching off to the left promising a more direct route to the top. I'm prepared for some bushwhacking. With only my parka pockets loaded with snacks and a small canteen of water, I am relatively unencumbered. The tracks leading into the forest, are overgrown with brush and even small trees. Maybe I only imagine them. Nevertheless, the decision is easy. The blacktop doesn't even rate a backward glance. The Jeep trail, imagined or not, has captured me and with only a few steps I am pulled instantly into the quiet magic of the woodland realm.

Yes, there is a trail. Away from the smaller growth along the road's edge, larger trees have blocked much of the undergrowth, and the trail is more pronounced, making progress easy and inspiring an overrated confidence in my intuition. I am able to set a good start pace – that easy rhythm of thick-soled hiking boots plopping on the steep incline. At the top of the first rise, more signs of man. Tree stumps fashioned many years ago by hand-axe it appears, rather than chainsaw. The stumps are weathered and worn gray, but they still stand stubbornly resisting final decay. With or without their trunks and treetops, the roots remain for their predetermined length of days.

Man would logically choose this site. It affords a commanding view of the mountain side and a glimpse of the lake far beyond. But the stumps have outlasted whatever structure they sacrificed their wood to build. There is no evidence of a foundation – only a large depression – but no other signs of man's intrusion.

The signs of spring in this wood are subtle. No wild, exuberant, splash of spring color. Under the shadow canopy of pines and hardwood, only small violets announce the promise of spring. They are shy, of course, and not very prolific. They require a patient eye, but reward persistence with a glimpse of their sweet and simple form.

Over the rounded crust of Bear Mountain, there is a gentle downward slope to the trail. No switchbacks required. The trail bends around a steeper proportion of the mountain formed by an outcropping of rocks. Two large tree trunks have fallen across the trail and provide an inviting rest stop. Without the noise of my own footsteps, I can enjoy quiet wood sounds. The trail is low here with slopes rising on both sides. Private and lonely hillsides completely surround me, look down upon me, accept my need to be here, and share with me ancient wisdoms. With eyes closed and head back, I feel profoundly at one with my surroundings. The intensity of my feeling seems to match the power of the place. I find in this one intense moment what I continually seek on these wanderings: solitude, complete, total and soul-filling.

TIMELESS

I sit in a meadow, a brown boggy mat covered now with a late spring showing of marsh marigolds – pale yellow and white in tight bunches. Neither shy nor tentative, these flowers jut straight out of the ground with thick sturdy stems and uncomplicated blossoms formed by a halo of petals, daisy-like around a central burst of small golden spikes.

I sit alone in the brown bog meadow, having tiptoed on clumps of dead grass to stay above the ooze. I have climbed over a large spruce trunk with swirls of its gray twisting roots exposed. I sit on the rotting trunk of that fallen spruce, grateful for the dry bench it provides. No other trees invade the meadow; instead, they stand guard along the perimeter as if protecting a holy place. The rush of snow-melt water in the stream at my feet, provides the only feeling of urgency. It flows like a liquid brown snake forming "S" shapes with its fluid body and hissing in a low breathless rumble.

I have come to sit along among simple flowers on a simple day. Above, a quiet and unhurried sun also sits alone in its cloudless blue sky pierced by mountain peaks.

There is a strong sense of place here, but not of time as if there had to be an adjustment between the two; more of one means less of the other. The stream, the evergreen forest, the meadow and its flowers all are connected with defined relationships to each other. These elements and their specific arrangements under the bright vault of sky provides a strong and sure feeling that I know unquestionably where I am, and that through the experience of this day, I will come to know this place as distinctive from all other others.

I have added myself to the landscape, and according to its rules, I have concentrated on my connections to the elements, putting aside considerations of time. The present is all that remains.

With the disencumbering of time, comes an intensification of spatial detail. The timelessness creates a magnifying halo around me through which every separate object is brought sharply to focus. Each flower, each tree becomes individualized, showing unusual markings, unique forms and shapes. Single specimens begin to stand out from the group, as if they have answered some silent rollcall; this one, and now the next, and the next being recognized as the separate entities that they are. It appears they have somehow straightened themselves or shifted slightly to catch my attention. Classifications of group, population, species seem trivial and insulting. It is the appreciation of one – just one, separate from the rest – that permits the intense inspection of all of its particulars: the overall height and width of the tree, the absolute number of branches attached to its main trunk, the distance of the first limb from the ground, the number of branch points in each limb, the distance between the tree and its nearest neighbors. These must be important items to a tree just as details of our own individual bodies are important to us. We fuss over the peculiarities of our foot size, the length of our eyelashes. We take delight in some small dimple that inexplicably gives sensual pleasure to our partner. We marvel over our own distinctive features reflected in miniature on our children's faces. The details are all important; we treasure them.

I am overwhelmed by the detail in just this one small meadow, detail usually ignored in the interest of time. Time robs us of this richness, of the diversity and profundity inherit in all living matter. I feel a special contentment as I sit alone and map the branching of the evergreens at the edge of the forest and the multitude of flowers on the meadow, as if reclaiming some forgotten rite, ancient and important. It takes time – more precisely it requires the disengagement from time – to engage in this practice. But we should remind ourselves that the world does not operate according to categories or other inventions of convenience which emerge from the human mind; it does not calculate probabilities or averages; rather, it comes with a blend of countless single events, each involving some individual detail. We live according to the interactions of minutia.

I claim myself as an individual as well. I assert my unique character, my small singleness, and make introductions to the great diversities of nature that reside here. I have broken free of time and sit simply in the middle of the mountain meadow without directed intent, other than a hope for acceptance, a hope to emerge as does a tree from the forest, as a separate living being imperfect, complete in myself; one who seeks to dwell near holy places with the confidence and clear purpose of spring flowers.

DOWN VALLEY GUIDE

The view down valley is still here
Today draped in summer green and yellow-browns
Broad-shouldered mountains
Point the way to far vistas
You can still see to the future

Our years together began with this view
But softened then by deep snow
A sweeping, sculpted path that called
And we answered:
Nothing else matters

With winter coat lost to the season
Scars appear in the landscape
The sharpness of the ridges and buttes,
And a different kind of textured beauty
In aspen groves and alpine meadow

I suppose we are also more textured now
And the richer for it
You, still the guide and me, the muse
I follow with you down valley
Still.

FROG MEDICINE

The rain is called forth with the voice of a little green tree frog; a pure, insistent note, offered in repetition. The call is full of certainty, a deep knowledge that the answer will come, the offering will be returned. He sings in rain-rhythm staccato, to punctuate the song's message. It blends with a hundred other forest sounds: wind whispering; the rush of water gliding not quite silently over rocks and against stream beds; the swish of leaves. His music originates in low-land places, builds, interlocks, rises layer upon layer until the heavens respond with a little shiver, a twitch, a soft distant rumble. He sings the song only he knows, with a confidence born of knowing the way things must be. There is laughter in the song; there are sounds of his passion. He closes his eyes, tilts back his head and hums his dreams. So powerful is this, that the mists begin to rise, wispy at first, swirling in among the stars; then thickening to hide the slice of new moon. Another shudder fills the sky, shakes the stars in a thundering reply. A dark cloud-brew pours along the horizon, sprinkling the path of the song with lightning leaps. The sky breaks. His song reaches up and pulls down the first clear drop that splashes at the edge of my eye, a star's tear. It hangs for a moment, then spills over the ridge of my cheek to rivulet across the flat plain of my face and reach the lower jaw line. It writes a transparent scar – maybe it is a transparent line of healing rather than a scar. Or maybe a fracture line, a crevasse that cracks through superficial layers. Maybe a cleansing line, an erasure of the deadening layers and pretense. I do not have time to think this through before the second drop, then a third and another – ten all at once splash down. His song reaches to touch my chin and tilt it toward the rain. As I am turned full-faced skyward, a song is pulled from my own throat and I, too, sing with the rain. Come clean, we all sing together, he and I and the rain. Come clean.

GIFTS OF AUTUMN

Sky Top Farm is in the arms of fall now. We have watched it come at a slow and pleasing pace, watched the color of the leaves change by degree, and felt a chill in the air turn sharper with each sunrise. The Juncos arrived this weekend, and now line the country road we run on. The great blue heron, fishing the pond south of the cottage, shared his/her favorite fishing spot with us but so far all we've had are strikes. Robins cover the open pasture late in the evening, sharing ground with a meadowlark who sings so sweetly. We rescued a sparrow hawk from the barn the other night. She was trapped there, flying in confusion among the dark rafters. She flew very close to me, almost touched my head, but left in such a hurry she didn't even say thank you. Red-tail hawks soar on sunny days and turkey vultures circle close, especially on weekends when the deer hunters fill their gut buckets with animal entrails. The tree where they dress the deer (isn't that a strange use of the word, "dress"?) is near the cottage, so we have a "hidden-spectator" view of the ritual: men encircling the kill, helping each other, talking with quiet laughter that seems a little subdued. I sense a reverence in the way they stand and watch; maybe it's my own reverence that I feel. We will hunt deer after Thanksgiving and hopefully make our own offering to the gut bucket vultures. So far I am content with road-kill (a fat woodchuck with a skull too smashed to keep, but we managed to extract the two impressive bottom incisors), windshield-kill (a beautiful question-mark butterfly kamikazed on my wiper blade) and screen-kill (a sharp shinned hawk, long dead and desiccated on the floor of a screened picnic area during our trip to the Massanutten Mountains – a fairly intact skull, tail feathers, talons, and sharp-shins harvested). The basket on the kitchen table in town holds a Sky Top harvest of hickory nuts, dried Queen Anne's lace, hand-picked cotton balls and my favorite: large leathery seedpods from a black locust tree. The only sort-of gross thing in the collection (but very special) are persimmon seeds that I gathered from dry coyote scat along the dirt roads of the farm. Sacred seeds blessed by the wild clan who sings close to the cottage at night. Unfortunately, with the cooler weather we hear them less often because their songs are lost in the hum of the heat pump motor.

Rusty has found another remarkable book, *River Walking* by Kathleen Dean Moore.* My only complaint so far after two chapters, is that Moore may worry a little too much about things, such as Emerson's warning: "For everything you gain, you lose something."** My view is, why bother yourself trying to keep tally? In Moore's terminology I am a "gulper," not limiting myself to tiny sips of happiness just to fool the gods who ration out equal amounts of pain and joy. Without doubt this fall is a season of joy-full gluttony for me.

*Kathleen Dean Moore, *River Walking: Reflections on Moving Water*, 1995.

**Ralph Waldo Emerson, "Compensation," 1841.

SKY TOP FAREWELL

From fifteen feet away, I watch a big otter taking a sand bath near the edge of a pond. He is sleek, like a thick brown muscle pointed at both ends. His head is similar to a seal's: two large circular nares at the end of his sharp snout, tiny half-circle ears. He has web feet and a rich brown pelt that is pale under the place where his chin would be, if he had one. He is one continuous flowing line from nose to tail tip. He appears to be strangely fastidious. He carefully rubs himself all over with the sand: back and belly and under his legs, scratching and licking here and there. A good grooming job.

A loud splat on the surface of the little pond quickly gets the otter's attention and mine. We both turned to see a Belted Kingfisher scooping up a frog and giving his distinctive rattle-like call as if to brag about it.

I am on a morning hike through the woods – the back way to the bottoms by the old tenant houses, along deer trails in the woods, bordering the north edge of the fields to the small spring-fed stream near the "kissing trees." Three deer were bedded down there; they bounded off when I got too close. I followed their path across the open field near the middle deer stand and re-entered the south woods. Following the high ridge to the Texas place, I continued east to the false catfish pond, which is home to the well-groomed otter. I have come to Sky Top Farm this week to fulfill an assignment. I have to explain to this place and to myself that I will not be living here anymore. My husband and I are moving to our new home at Hawk Ridge next month. It is as if we are stepping across a small stream from one side of the bank to the other, each side is wonderful in its own right; each has paths that are part of our lives' journey. We have reached the crossing over point.

I have left the pond and turned north by the East deer stand, the tank, and come along the road below the corn feeder. I am lucky to have found the same comfortable log where I sat two years ago while deer hunting. The breeze here is terrific; poison ivy minimal. A blue jay, a chickadee call. Leaves are just beginning to fall. One here, one there, each making a small rustling sound. I will remember how this feels – dwarfed by the size of Sky Top's 2,400 acres. I will see no one, talk to no one in person for these two days unless I choose to visit Billy and Mary Nell who live down the road. No deer hunters are here this weekend; I will have the entire place to myself. I could continue hiking for the rest of the day, never leave the property, and have complete solitude.

I will remember the lay of this land – hills and deep gullies, hardwood trees, thick mats of leaves in the woods, the open fields with curving borders dictated by the hilly terrain. The secret places – deserted barns with owl pellets, woodchuck dens, robin's nests; old stables with saddles – unused for years – still draped over the rails; a hanger storing an ultralight airplane and jet skis; twelve lakes (did we find them all?); a huge tree split from tip-top to root by lightning; the deer trails that take you everywhere you want to go on the place; fungus of every description growing on fallen logs. We have seen over a hundred species of birds (many added to my life list): hawks soaring, fly catchers darting out for insects, graceful landings of the great blue heron, flocks of high-flying geese in wispy line formation, the loop-de-loop flight of woodpeckers; the ceaseless fluttering about of warblers, the miniature dive-bomb antics of hummingbirds. We have taken the time to watch the show.

We have grown vegetables, picked blackberries, blueberries, and pears; feasted on quail, bluegills, bass, dove and venison from this place. A hundred evenings spent on the front porch watching fish rise on the lake and deer browse on the ridge across the road; listening to barred owls, screech owls, wood thrushes, whip-poor-wills, and chuck wills widows; peeking around the corner of the porch to make sure we didn't miss the sunset. Counting stars, counting blessings.

It is not sadness I feel now – not exactly that. I feel full of memories. I have gathered them up around me, as if I had them in my apron pulled up by the hem. My arms are full of these trailing memories, the map I drew of Sky Top Farm is tucked under my arm, the treasured deer skull with thick antlers we found is in my one free hand. It is good that the soil of Sky Top is not living skin; that the roads and deer trails and sand creeks are not veins and arteries; that the wind sound emanating from the trees is not an actual voice; that a beating heart does not reside at the base of the black locust tree at Four Point Place; that the one yellow brick in the walkway to the cottage, does not cover the secret opening to a soul living there. Even though it feels like that to me, I know it is not true. If it were, then there would be sadness. If Sky Top were capable of knowing how much we loved it and how unlikely it is that anyone who comes after it us will love it as we do; this would be a sad time.

The days we have been here surround us like a halo. We have left our mark on Sky Top and it has marked us. Even if deer trails and lakes and brick cottages are not living things, they do create a spirit, an aura that invests me and the man I was meant to love. In that way he and I embody the spirit and it gains a life through us. With this definition I am able to say my goodbye, finish my hike, and head to the big lake behind the Oates' house to try one last time to catch a decent size bass.

BOG LAKE BLESSING

Have you ever seen a solitary loon
Fly so close overhead that you could feel
The wind from its slow, deliberate wing beats?

And did an elongated sunset play itself out
Over a smooth-surfaced lake
Reflecting the closing progression of white – gold – pink?

Was the evening stillness split apart
By the loon's trembling call
In affirmation of connection to a sure and lovely wildness?

Did you stand very still and let this moment
Explain your own belongingness
To the dimming sky, the quiet lake, the cry of the loon?

TRANSITION

Emerging from the lake, goose-pimply and shivering
Escaping chilled water barely swimmable
The flash of bright sunlight hits bare skin
Spreading elemental warmth
Instantly mellowing.

Or
Standing with eyes closed
Head tilted back in a sweltering sigh of exhaustion
Feeling the first cool drops of rain
Rinse away sweat and grime
Into small rivulets
Flowing across your skin.

Or
Stepping suddenly from the city's cacophony
into the stillness of the Maine woods
Granite boulders enthroned on their cliffs and ledges
Uncaringly oblivious to passing moments
Of pretentious importance elsewhere.

Human experience is segmented into bits of time
Apportioned by the ticks of the clock
Each anchored to coordinates of the surround.
Keeping time and hopping stone-to-stone in our journey,
We are focused on where next to land.

But what of the time spent mid-air

Leaping in transition
Floating in a brief instant of flight?
I long, impossibly, to stop
Precisely there

Freeze-framed
Untethered to any single place
Suspended between before and after
Embracing the sum of all pasts, all futures
Quivering in the intense pleasure of change.

CREPUSCULAR

What is it about evening time
That feels so different?
Is it because we lose sight
And must turn to other sensations, other feelings?

Daylight gives us crisp edges
Definite positions for where things are.
We can know for certain
The attributes of things around us.

At dusk
Our surety pales
We are not quite so sure anymore
Where things are exactly

As if our minds begin to make up things that are missing
Did I just see a shadow move?
Reality and our grasp of it seems to melt a bit
Lose distinction

And are our emotions the same?
Do we feel more vulnerable as the light dims
As the bustle of the day quietens
And we become aware of questions, of unknowns?

This special time – crepuscular
A transition, a necessary acceptance of our changing place
A letting go of the known, the expected

And an embrace of the less certain

Confidence to do so
Is a special blessing
To accept and marvel at what is unknown
And slip easily into the black of night.

SHORELINE

A meeting place
Strange intertwining of two such different milieus
A defining border between solid ground
And the ever-changing ebb and flow of dancing waters

A tempting interplay
Enticing us in past eons
To leave the ocean's womb
To emerge and place our first timid steps on firmament.

It still calls to us
To sit at water's edge and look out across the shimmering surface
To the beyond
To past, to future, to all the awaiting possibilities.

BOOT HEAD RECOMMENDATIONS

There is a bench that occupies a magical spot at Boot Head Preserve. You might think a bench is ordinary, something to be expected along a Maine Coast Heritage Trail, but that is not the case. Along the dozens of Bold Coast trails my husband and I have hiked during our twenty-plus summers Down East, I can recall less than five benches – total. The Boot Head bench sits at the very precipice separating Boot Head from Brook Cove, a cataclysm of sorts – sheer cliffs that drop straight into the ocean, lining a deep cleft in the shoreline. At high tide and with rough seas, waves pound the granite, forcing their way into the narrow opening of the cleft. In a repeating cycle, each new wave adds to the previous, finally producing a powerful spume that is forced straight up in an explosive display. It can take several waves to complete the cycle. In our mind that is why the bench is placed precisely here. You sit quietly, start counting the waves and wait … wait – BOOM! The blow hole, as we call it, provides its chest-rattling roar and powerful thrust of tons of seawater several stories high. We have stayed perched on the bench through many repeating cycles, all the while feeling tempted to peak over the edge and watch the event, perhaps get close enough to be anointed with drops of the spume. Not recommended.

Nature's waterworks are among many attractions at Boot Head. Less than a half-mile on the trail from the parking lot, we have stopped to marvel at the extensive heath, with rare apple berry and pitcher plants close enough to get good photographs from the trail. We have followed the trail loop to Boot Cove and its small, half-moon beach, then out along the rugged cliffs to reach the bench, hoping for favorable seas that will "blow the hole." All highly recommended.

ALASKAN FEAST

A small gathering of eagles
Calling to each other with thin, sharp whistles
Signaling anticipated abundance

A tumble of dark leafy greens piled along the riverbank
Scrub brush and trees splashed with golden spice
Hinting of fall

Crystal blue-green water from glacial melt
Aquavit flowing clear with the wind
Curling the lips of waves with white froth

Preparations for a banquet table
Served beneath a silver grey dome of clouds
Opening to a Chinook arch of blue innocence.

Bears and birds, even trout
Gorge on glistening bounty of spawn and spent salmon bodies
Ripening to rich reds reflected in the autumn tundra

An offering of sustenance on a monumental scale
A magically balanced cycle of death, scavenge and emergence of
new life
Drifting as small orange orbs decorating the riverbed.

BIRDING WEST TEXAS

The Chisos Mountains, an ancient seabed risen from the watery
    depths
Displaying the incomprehensibly slow pace of geological time
Creating an austere, stark landscape, inhospitable in appearance
At the same time, magnificent on the grandest scale.

The volcanic rust red of flat-topped mesas
Seared by the West Texas heat and wind
Palisades banded by ocean sedimentary layers
Survivors of briny seas, now decorating the desert landscape.

Somewhat startling is the number of birds
Happy to call this home.
Among them, a rare and thus sought-after species
The Colima warbler.

A tiny smidge of a bird, dull green with a dusting of red on its head
A simple, short warbling trill for a song and an almost inaudible chip
    for a call note
Elusive and very discerning in its nesting habits
In all of America, they choose only one small sliver of a 7,000-foot
    Chisos mountain top on which to build their nests.

The quest for the Colima requires an "up-at-4am-to-get-to-the-eight-
    mile-hiking-trail-in-the-pre-dawn-darkness" effort in mid-
    May
Lots of water bottles, flashlights, binoculars, hiking boots suitable for
    rock scrambling

And importantly, a firm but difficult to explain, Zen-like
commitment.
Best to avoid stopping in the middle of the trek and asking yourself,
"Now, why am I doing this?"

Few people see a Colima,
One of the most challenging of all 700+ nesting species in this country
to add to your life list.
Our group of four family members recently paid homage to this
special little warbler,
Trekking over its desert breeding ground without sight or sound of
the bird,
In spite of heroic efforts from an accompanying birding-savant guide

Yet it is the quest itself that is most satisfying to the soul, not the
conquest.
While Colima is not Moby Dick or snow leopard,
It holds its special place, not caring one whit whether I ever glimpse
its small form.
It challenges us by its very existence – and that is quite enough to
celebrate.

Cardinals Singing

It seems like such a long time since I have written – last season, the winter past. It has been that long since I have shed the yoke and harness of pressing matters, matters of the other worlds. It feels as if I have just lifted the covers, pushed back quilts and slept-in sheets, to welcome a freshness, bright and clean, as it brushes against my face. It – meaning my place in the natural world – has been here all along, of course. I've been hiding from it and it has been invisible to me.

I am greatly relieved that the heartbeat of nature does not depend on me. Seasons cycle. Tides ebb and flow according to countless internal clocks, all ticking and sounding the wake-up alarm for earth's minions more or less at the "right" time. At least it's close enough, so that the life force can maintain its fragile foothold on earth's crusty mantel.

I hear a cardinal singing. Such a flatterer. "Pretty, pretty, pretty" it says, and the early spring winds strum the trees in rhythm with the refrain.

I'm settling in now, feeling right at home although it seems I misunderstood the cardinal. He has just dive-bombed through this stand of trees on the tail of an equally brilliant male competitor who probably invaded his territory. Now he's singing again: a victor's song and a challenge to all comers. Female cardinals – two of them on the top branches of the tallest tree – are preening. Maybe it's a love song, too.

It is a mystery to me how Dillard* and Thoreau,** both possessed of the deepest understandings of woods and wild creatures, could follow those understandings back (backwards from my perspective) to philosophy and human spiritualism. If ever I am to understand anything at all about the cool spring earth I sit upon or the young tree that is a cushion for my back or the glorious songs of birds, I doubt they will lead me to human philosophy. I fear they will take me instead to a separate place and I will not speak to humans often.

*Annie Dillard, *The Writing Life*, 1989.
**Henry David Thoreau, *Walden*, 1854.

FLOWERY TUFTS ON HIGH MEADOW*

Not a scythe, but a Massey Ferguson tractor
Model 231
Fire engine red
A six-foot-wide bush hog attached

My hard-working companion of the past twenty+ years
Affords me sweet delight
On sunny afternoons in High Meadow
Late summer; butterfly time

The intent of the mowing
Is to maintain a patchwork of woods and open fields
Creating edges that are a special meeting place all wildlife seem to seek
Borders between open foraging and shelter near-by

The swath of the bush hog can, with care
Create pleasing patterns in the mown grass
Long parallel lines that reach straight or slightly curved
Across the meadows to the forest edge

Yet artistic perfection is not the only goal
Other needs can override
In particular, the irregular patches of flowering New York Iron Weed
Attracting hundreds of butterflies deserving sanctuary

Granting them wide birth, disrupts the otherwise perfect pattern
Arching the tractor's path around them, creates a bouquet

Dense blossoms of deep purple
Animated by swarms of butterflies: tiger swallowtails, cloudless
sulfurs, monarchs

With the turn of the steering wheel,
I give them my gift
With the silent shimmering of their wings
They give me theirs

*In honor of Robert Frost's "A Tuft of Flowers," 1915.

THE HAMMOCK

A writer friend of mine enjoys a certain degree of success because she begins her novels with some world-wide crisis, and then brings it down to human scale using details of each character's life. It's like looking through a telescope. Monumental events are more comprehensible this way, taken bit by bit, and then fit into the human scheme of daily experiences.

The things I have to say as a writer are a very different sort. I am fascinated by minutia – those wonderful little episodes in our lives, brief glances of ordinary things that are common in one sense, but nonetheless of great value. The teeny tiny things – a little red spider; a single oak leaf, smooth and leathery brown; one million individual tuffs of green moss covering a fallen log like a burial raiment. All it takes is a quiet walk in the woods to set up my mental microscope and bring these small wonders into focus.

How could I have missed them before? Aren't these blessings of nature always around us? Are we so sanitized and organized that the wildness of nature has no place in our lives? It seems at times that we have sealed ourselves up in our working/living nests surrounded by our assortment of imagined necessities, and that we have sealed nature out. We are comfortable and comatose in tombs we have fashioned with our own thoughts. Our geometric minds want square corners, clean lines, lifeless white forms that are refined and everlasting. We shun the chaotic curve and flow of nature, its cycle of change, its meandering energies that move from wind and sun to restless leaves and rabbits, to foxes and eagles and weasels who eat the young. It is unkempt. It glories in a freedom we seem to have laid aside.

I sleep in a hammock whenever I can – under the trees in my

mother's backyard, on the patio of a hotel room during a business trip, in the wilderness on backpacking tracks alone or with friends or family. The device itself is a wonderful design, very lightweight. The crisscrossing strings are strong and durable. A simple slipknot secures it to trees, patio, awnings, posts of any kind. In a pinch, a single tree will suffice, with both ends of the hammock tied together and me sleeping curled up like a rock in a slingshot.

The hammock gives me the moon and the stars; the evening winds touched by sweet green fragrances that leaves release only at night. It gives me strange sounds in the dark that frighten me and test my courage. It gives me a gentle rocking that lasts for several minutes every time I shift position. It gives me rude wake-up calls from hummingbird so close that the whir from their wings sounds loud in my ears. These treasures are all around us. They are always here. Every single morning of our lives begins with a gentle pre-dawn light that unfolds so very slowly that it cannot be paced. To sit in the stillness before dawn and ever so quietly receive the first light of day – this simple thing is so powerful that it can change your life.

Friends are shocked when they learned that I often sleep alone in the woods. They tell me that the idea is so foreign to their everyday experiences that it seems bizarre, downright crazy. I'm always saddened by this response. The risk of bodily harm is much less than that associated with an automobile trip during any holiday season. And the small inconveniences are no more than a little pinch from nature just to assure me that I am still awake and alive and that I can still feel.

The woods are not quiet or dark at night. They are a place with starlight that is lost in light-polluted cities. The deep darkness of a forest intensifies the moonlight – queer, opalescent, polarized. We, like all animals, feel differently in its glow, and our bodies maintain a natural rhythm according to its changing phases. It is a celestial timepiece that paces our own reproduction, our own birth.

Sounds of the night also have a special quality. Nothing is quite so eerie as hearing something in the dark, especially when there is no assuring light switch close at hand. Part of the problem is that we are so unaccustomed to animal sounds in the wild that we haven't a clue whether that ghastly sound we just heard was made by a little green tree frog trying to attract a mate, a hungry grizzly on the prowl, or a raccoon fussing at her babies. In our ignorance, it is hard to simply accept these sounds of unknown origin as readily as we do the drone of traffic outside our bedroom windows that drugs us to sleep every night. It is a chilling thought that we might actually be in physical danger from something we can't *see*. Regardless of how often we acknowledge that the statistical probability of being attacked by a wild animal in the middle of the night is vanishingly small, it doesn't seem to ease the concern.

The first few nights on a long trek, I doze fretfully, awakening to each strange sound through the midnight hours. I sleep only by force of will, but with the very first blanch of dawn when darkness has eased by only one degree of gray, the tension is released to a sense of calm and commitment unmatched by any other experience I have known. I am so pleased with myself for having toughed it out and so grateful to the wilderness for this intense pleasure, that I fall into a heavenly sleep, extraordinarily sweet and refreshing. When I finally wake to the full morning, I am loathe to leave my suspended nest in spite of the urging of a new day

It puzzles me that these experiences are so meaningful; that they speak so directly to my sense of well-being. They are not out of the ordinary, but rather they are intensely <u>in</u> the ordinary; usually private moments that are mundane, mundanely trivial, dwarfed in the absurd by important issues of the world. But they bring real joy, intensified and pure as laser light. They bring a chuckle, a squeal, a sigh, a shiver; they bring me to life.

THE RAVEN

There is a mountain pass in Colorado that reaches over a long and sinuous ridge, offering views to the deep valleys on either side. One view stretches to the plains of the Southeast; the other to northern peaks of indescribable beauty. The early November snow lies over the pass, thick and freshly poured. Its delicate softness hugs the trunks of blue spruce and pine, and flows over the scattered rocks to smooth their sharp edges.

The air is thin and sweet with no wind to disturb it. It is so quiet that I can hear the wing beats of a blue-black raven as it flies overhead. Without warning, my spirit leaps up to hitch a ride and soar with the bird, sharing its vision of this whitened world. I am instantly clutched in ecstasy, stunned and breathless as we sweep the highest peaks together. I am dangling mid-air with limbs trembling and my heart so full it would seem ready to explode. I am carried with dazzling speed even higher, up toward the bed of ice crystals in the sun's aurora.

My attachment with the real world is stretched thin to the breaking point, and it is all I can bear to hold on to it while still remaining in the raven's grasp. It is more than I can bare, and a bond must break. Slowly I slip from the curved talons and fall to the snow in a jarring sadness.

I sit quietly now, wrapped in the snowy blanket under empty skies. It was after all, a tenuous joining of my heart to that of the raven.

This day has told me that you can't always get all that you want. Perhaps not. But sometimes the mountain gods smile, albeit briefly, and present you with great gifts. I will take the gifts with gratitude, though I cannot comprehend them. In time, I will descend the mountain pass, leaving love hearts drawn in the snow.

ATLANTIC PUFFIN

A football with wings a bit too small
An elongated clown's nose in riotous orange
Throws itself off a ledge, chest-first
Flapping furiously to launch its circus act

COMMON LOON

Two of this ancient lineage
Protecting their chick, always within a feather's touch
Feeding in tandem, one watching while the other seeks prey
Haunting the lake with eerie duets sung at midnight

DUNLIN

For winter, a dull grey overcoat
Hanging out with buddies
Dense flocks in aerobatic display

For spring, an apron-tuxedo
Black, over bright white belly
Nesting time announced
With a gleam in its dark button eye

LEAST SANDPIPER

Fidgeting shorelings
Multitudinous to say the least

Each tiny jewel perched
On exquisite yellow-green legs

Slender bill stitching through the sand
Folded wings dappled in soft taupe and tan

TRUTH COMES TO LIGHT

In the early predawn shadows, I saw a woman mountain. She was laid out above the western shore of Cedar Lake and she was staring at the fading stars. Her profile appeared in perfect proportion. The hair flowing back from her head was thick and long with streaks of silver; curving tangled locks which twisted and tumbled down to reach the water's edge.

The eastern slopes had blocked the sunrise, acting like a massive sundial of sorts that slowly measured out the light of day. Their peaks directed the sun's first rays toward the woman's upturn face. As the bright leading edge of sun/shadow began to descend, I witnessed a transformation. The light seemed to undressed her, peeling off layers, not of clothes, but of skin and flesh to expose the bony details and hard structures beneath the feminine form: structures that had lain hidden with the woman through the night. The shape of her eyelashes became a crest of three slender trees, silhouetted against the morning sky. Her chiseled cheekbones and brow were set into high bluffs of gray and brown. Her breasts formed heavily forested slopes with sinuous outlines. Her tresses flowed into gullies, across stone ledges and rock slides; her gray-white streaks of hair settled into patches of snow. The sun burned through her soft shadow figure as she gave herself over to soil and rocks in the hard light of day

When the sun light finally spread across the lake and reached my hammock on the eastern shore, I arose, and the woman mountain was gone. I could not find her image again, though I looked throughout the day.

LOVE OF FAMILY AND FRIENDS

Moments With My Father

The ultimate problem-solver, driven to find possible solutions for his
    family, for his community
"If you know Cliff, you know he will work"
Became his political/community service motto

Private moments shared with Daddy were rare
In their frequency, yes,
But they created deep meaning in my life

He "taught" me how to drive when I was five:
Sat me on top a pillow in the red Willis Jeep driver's seat
So I would be able to see over the steering wheel;

Turned on the "governor," set it for five miles per hour
Instructed me to turn the steering wheel to keep the jeep pointed
    directly behind the hay-baler
Which in turn was hitched to a tractor (driven by my seven-year-old
    brother)

Hitched to "my" jeep was a large flat trailer
Several high school boys had been hired to walk behind the baler
Loading hay bales on the trailer as soon as they were ejected from
    the baler.

I have loved driving ever since; applying the lessons learned to life's
    many challenges:
Accepting the responsibility it requires
Embracing the feeling of competence generated by completing a
    difficult task

In similar manner, Daddy "taught" me how to squirrel hunt
(Showed me how to load my BB gun, sat me under a tree and wished me luck as he left)
I have loved the taste of fried squirrel ever since

When I was seven, he took me along on a late-night, deep-woods coon hunt with neighbors
They had coon dogs who howled eerily under a full moon when the coon had been treed
Which may explain part of my fascination with camping/sleeping outdoors on moonlit nights

When I was in high school, he allowed me to use our tractor to mow our neighbors' fields
For which I was paid a relatively handsome sum
And I got to keep all the money - much to my brother's chagrin

Daddy and I shared a deep bond with tractors
Growing up in the age of mules and oxen, he thought the invention of the tractor was a treasured gift
I still mow pastures with my red Massey Ferguson tractor – a 60th birthday present from my understanding husband

Daddy also established the "tractor bond" with my son Spencer on his third birthday
Placing Spence in the tractor seat (governor set at slow speed)
Then walking alongside as the delighted three-year-old steered the tractor solo.

Only two years later, at age sixty-two, Daddy was driving that same tractor (solo)
When a heart attack took his life
He left us much too early
But he left us with gifts
With rare perspectives of our home place
And with challenging legacies to fulfill

Queen of Make-Do

The youngest child; the only daughter
A favored and pampered position
In a South Louisiana community – French/Acadian/a bit of German
Tightly woven, intertwined,
Lots of cousin marriages, including that of her own parents

Life was play/work/visiting relatives on Bayou Blue – a joie-de-vivre.
As was the custom, her family adopted a younger cousin
When his own mother passed away.
The arms of the Matherne family extended to embrace the new
younger "brother."
Now, as older "sister," she had to relinquish her princess position.

Perhaps this was her first lesson in "make-do"
To balance wants/needs/expectations
Without losing the deep perspective of:
How do I want to live my life?

Themes reflected in the only poem I knew her to write (but never
finish), entitled:
"Who Am I?"
Followed by a list of her roles, responsibilities
Including what gave her the deepest joy

Items I would add to this last category:
Finding four-leaf clovers
Being ever-ready for the next adventure my father had in mind
Creating yummy meals for five kids in the back of the car as Daddy

drove to who-knows-where
(My favorite: Open-faced cheese sandwiches perfectly melted by the sun coming through the back window of the 1956 Chrysler)

She approached cleaning house and other duties in the same manner
Having my older brother, Robert, and me sit on a large polishing rug
While she pulled us around the living room to complete the chore of waxing and polishing the floors
As she side-stepped and swayed to rhythms of Jimmy Dorsey's jazz band wafting from the 78 record player

My father died of a sudden heart attack at the ridiculously young age of sixty-two.
My mother coped, with a resilience that comes from such strong family ties
She arranged for his burial as he had requested
His isolated grave atop a hill overlooking the lake he loved
Then, after two years, she had him disinterred and moved eighteen miles to the town graveyard
Where she could more readily visit graveside
And where she could more readily be buried beside him, when her time came.

But she was angry
Very angry about loss of the golden years with Cliff

The only antidote I found to offer was
A two-week car-trip to see her myriad cousins in
Houma, Bayou Blue, Thibodeaux, Big Mamou, Raceland
From New Orleans to the Atchafalaya Basin, we made the rounds,
Each stop – an emersion, a cleansing, a reaffirmation

The interludes with just the two of us in the car driving to the next cousin's house
Were like episodes of her favorite radio soap opera, "Helen Trent"

An opportunity for her to articulate so many things about life with
Cliff Ammons
Some – rather embarrassing even to a daughter
But the fully detailed, no-holds-barred story, was indeed a healing
celebration for both of us.

Mamie continued as queen of make-do for the twenty-five years she
lived after Daddy died
The same basic themes
But including more travels with friends and family
Always, always a unique influence on all those she touched

She persuaded my reluctant ten-year-old son to take his daily bath
(no shower available) when staying with her for a week
Her successful approach: add floating, lit candles to the water in the
bathtub
He loved it.

And when Mamie died
Her battle with cancer was a dance to the end
Her last days with hospice in her home
Were spent with her children and friends gathered at her bedside
Sharing stories, reliving important moments
Her last words still ring in my ears:
"I love you; I love you"

Sweet Barter

On my grandfather's farm
The end-of-summer air
Dripped with sticky sweetness
Steaming from black boiling cauldrons
Thin yellow juice of the sugar cane
Thickening to amber.

Begun as fragile sprigs
Piercing the red dirt hills
Demanding cruel, aching work
To grow explosively
Through sweltering oven-hot days.

Ending with the hack and slash of machetes
Fallen canes, jack-strawed
Curiously purple and pale green
Gathered up in muscular arms and brought to the syrup mill.

The crushing stone, in its frightening anger
Consumed the canes
Fed by quick agile men
Risking repeat of legendary maiming.

The stone was turned in its trough by a cussed mule
Responding singularly to the seeming cruelty of my grandfather
A clever, common-sense man
Who contrived the completely self-sufficient operation.

This yearly ritual came
With friends and kin gathering as bees to the final distillate
The young ones, myself among them,
Corralled at a protective distance
But jostling for the front line
To see tantalizing dangers played out for real.

With frightened fascination,
We crunched samples of peeled and sliced cane between our teeth
Squeezing out a sugary antidote
Against scary scalding nightmares that might come.

The final molten essence
Poured in a viscous stream
Into rows of waiting tin buckets
Each sealed with a flat-lipped top
A southern currency to be sold, bartered and gifted.

Sixty years hence,
It all bubbles to the edge of my consciousness
As liquid memory
A single succulent drop on the tongue.

Holding On

I guess I was about eight in that picture; one of three kids with a fish SO BIG it took all of us to hold it up dangling from a pole hoisted on one end by my older brother Robert, who balanced it with much effort, elbow to his waist and a bit of swagger in his wide stance, wearing his much prized high-topped tennis shoes and trying to appear aloof; baby brother Kenneth in the middle, arms reaching far above his head stretching his suspenders and his little fists pushing up resolutely on the pole like a weight lifter; then the fish; then barefoot me in my most favorite blue jean skirt, leaning very close to the fish, nearly touching the gaping mouth while awkwardly balancing my end of the pole; the three of us holding on together, perhaps trying to hold on still; but looking smug about it then, showing off the catfish Daddy caught in Lanana Creek and brought home like a trophy, as if we had – at least for that one moment – the whole of our world in a shared grasp.

A Trip to Shut Eye Road

I am parked in my van near a bend in Shut Eye Road just on the outskirts of my hometown. I have always liked the word, "outskirts," and I like the place – just on the fringe, the border between home and the destinations beyond, a place that offers options: you cross here to leave home or to return to its safety. But Shut Eye is even better than that. It is a low-down, nasty place; absolutely off limits to nice girls and especially off limits to me as daughter of an elected town official. Thus, I have waited a respectable seven and one-half years after his death before venturing down this forbidden path. It is curious how a road gets a bad reputation just by being a place where certain things may have happened. Shut Eye, deserving or not, is at the bottom of the list of places you want to be seen. The town's stigma placed here is hard to remove. Even now, try as I might, I cannot turn this place into just another pleasant country lane in my mind. It is low land, flat, unlike the gentle rolling hills that surround it. Shut Eye is narrow and winding with trees close in on both sides. The land is swampy, full of shallow ravines that hold standing water for a long time after it rains. The road runs roughly parallel to the main highway into town, but it forms a dark, curving course compared to the straight, clean line of the highway.

There are many branches off the main road, red-rutted, unwelcome looking paths that beckon slyly but only seem to offer misadventure. Assorted trash decorates most of these turnoffs: beer cans, wine and whiskey bottles, and various other memorabilia from high times on Shut Eye.

Dark clouds overhead lend a sinister backdrop for an old pickup truck that is approaching. As I feared, the driver is slowing down and his deeply weathered face turns slowly to keep his eyes locked on me as he nears in an obvious attempt to recognize who I might be. Yes, this is guilt I am feeling again, that familiar queasiness in the stomach. What was Cliff Ammons' daughter doing parked on Shut Eye Road? Well, maybe everyone will just consider me eccentric. Oh, I like that! Big city education can be blamed for doing damage to my Southern up-bringing. The thought is very comforting; it feels right. Isn't this a great way to go home again? To be accepted and loved by people you grew up with, people you care about in the deepest context even though you have lost all contemporary commonality?

"She's just a bit eccentric." That means no threat to the status quo, no confrontation; yet it offers a mechanism for acceptance even when I break their rules.

The old man looks a little confused by the grin spreading across my face, but as he drives by, with only the slightest hesitation, he lifts his right index finger off the steering wheel and points upward, the mime for "how y'all doing," known to all small town and even some big city Southerners. The dust settles as the truck passes, leaving Shut Eye Road to the grey sky, the van and a newly contented eccentric.

Have I Changed So Much?

I used to view my life as a smooth continuum, an unbroken line not necessarily driven by single purpose, but nevertheless moving in some general direction with only particulars of the journey dictated by opportunities of the moment. Now it seems there is a major rift, a fault line that has separated the phases of my life and formed a deep crevasse between them. The early years appear as a golden glow very near, very real and I am now unchanged in view and feelings from the young girl full of adolescent exuberance for the beauty and meaning of nature. I am still the tomboy of thirty-five years ago, lying tummy down and cheek pressed to the sweet pastureland on my father's farm. I delight now, as I would have then, in having found a small, shallow cave beneath a live oak that stands like a grandfather, offering a cozy hiding place for armadillo, raccoon or rabbit in the protection of its gnarled roots. As I finger the wondrous patterns of tree bark, it registers the same pleasure in the same familiar places of my brain known many years before. I delight in the scatter of leaves before me – some still in brilliant fall colors – others in beautiful browns and buffs, crisp and crunchy, layer upon layer. I know them, their familiar shapes and smells. There is a bond of deep knowledge that stretches across many years and reconnects the two separated parts of my life. This is who I am.

And who was I during the intervening years? More than a quarter of a century slipped pass and hindsight only provides indistinct images covered by fogy mist that hovers over the middle of my life. I feel I have just emerged from it as if newly awakened to my identity. The time of separation between the early years and the present is a mystery. I look back to see the young woman I was, moving through that time driven by happenstance as in a dream time.

I have little understanding of the period. I don't know how it felt.

If I reach back to those early years, I find that the last deeply felt memory I have is of an early morning race on wild horses through the pastures of a friend's dairy farm. The horses hadn't been ridden on a regular basis and they were skittish, unpredictable and given to tearing off in full gallop daringly close to the fences. My two college buddies and I finished the run as breathless as the ponies, gasping in the chill of the winter dawn, and blowing out steamy breaths that hung 'round our heads like a veil.

I remember the country breakfast of hot biscuits and cream gravy waiting for us. It had been prepared by the farmer's wife who was as wild as the horses, her black hair loose and flowing down her back untethered. She was also skittish, given to breaking away from the common life of the dairy to take drama classes at the college in town and to seek friends there among the passionately confused young women of the '60s.

The walls of her small kitchen were papered with clippings from the New York Times. Piles of books, stacked haphazardly in the kitchen, living room and hallway, formed a respectable library covering the classics and weighted toward philosophy and the arts. This remarkable woman had made her place in the world, filling it with an eclectic collection that included her own serious intellect, her rosy cheeked dairy farmer husband, a passel of kids, and friendships with country folk and young college rebels alike. We adored her; we pitied her; we admired her; we learned many things from her. She in return was sustained by our energy and adoration. We gave her a silver pin engraved with our three initials, ABC, for Ammons, Barbee and Casey. It matched the ones we wore, except hers was a smooth oval in shape; ours were silver bars with straight edges and sharp corners.

I have thought of her often, and my feelings for the wild horses, for her and for my college pals are the same now – the very same – as they were then. I hold on to all of them through memories newly bright and intensified by their absence from my consciousness of recent years. I hug them tightly in my arms and feel as though I have once again gripped the neck of the wild horses in full run. Like a weasel that has gone for the throat, I will not let go again until the end of the ride.

Charlotte

Charlotte has mesmerized me for weeks. I have watched, admired, and learned many things from this simple but wise creature. Her performance tonight so moved me, I was able to do little else but sit silently, staring as if I had been caught in her magnificent web.

Part of her allure is that she is so threatening, such a large spider with striped legs and markings on her thorax that resemble a menacing face. Her web measures a full two by three feet, strikingly beautiful with glistening threads so close, so evenly spaced, so pure in its design. Previously, we rarely saw Charlotte move. From our perch on the stairway, my son and I peered at her motionless form each morning, trying to reconstruct her unseen movements from the presence or absence of insects caught in the web; or from the damage or repair done to the web since our last sighting. The large window glass is the only barrier between our closely pressed faces and this terrifying but magnificent creature.

But tonight we saw Charlotte move. No longer the stone-quiet figure waiting, surrounded by her tightly stretched web skirt; tonight she moved, she circled, tumbled, quivered. She danced and the old web previously in disarray was transformed into a shining tapestry. Her web-spinning dance had intricate steps. The hind legs guided the silvery thread from her body. The front legs performed fluttering movements, weaving the thin lines into graceful patterns. The rhythm of the movements was a relentless beat, constantly repeating, pressing, as each completed circle became smaller, closer, tighter, moving toward the climactic center to end the dance.

I am moved by Charlotte's life, by her accomplishments, by her brave and solitary life as a resourceful, cunning, even treacherous hunter. The routine of her nights of work and her days of waiting, repetitious like the pattern of her web, contrast with the uncertainty of her success. Some days pass without a catch. More commonly, insects decorate the web: gray fuzzy knobs wrapped like mummies, a bounty set aside to be consumed later. I can't help but think that in relative terms, she is remarkably successful, a spider of substance, especially fortunate and wise to have found such a perfect spot for casting her sky-net. She is definitely someone to be reckoned with, to learn from.

I try to decipher the silvery messages she writes on the window each night. I try to grasp the meaning of her beautiful, enchanting dance; of her ruthless and deadly power. I am mesmerized by Charlotte, intellectually entangled in her simple snare, wrapped like a cocoon, still, motionless, separated from the world by her spinnings. She forces me to look inward, to the nature of things, to the heart of things, to my own rhythms, my own dance, to the messages I must write and cast out in my web of words. Then I must wait with Charlotte's patience to gently snare you.

The Wesley Pine

Snow blankets the blueberry barrens along the Wesley plateau.
The broad sweep of an almost tree-less horizon gathers to a deep blue edge
That cuts through the early paleness of the sky.
From some hidden nidus, sunrise erupts in an explosion of pink-orange flame, splayed out against the awakening heavens.

An improbably placed white pine is inked in against the rising sunrays
Its trunk is arrow straight.
How did it come to be there?
Seemingly unconcerned about the lack of neighbors, it stands alone,
Dominating the barrens and sheltering one small cabin hugged within its shadows

A tree, a cabin, a snowfall – in the tapestry of a Maine sunrise.
This speaks a prayer for the world.

SUMMER AT BOG LAKE

Consuming twenty-one crackers with
        cream cheese and Atlantic salmon
Discovering the two bullheads that swim
        near the dock
Delighting in the little brown bat skeleton, the polyphemus
        moth, the porpoise skull
Enduring the benevolent bullying of older cousins
Fishing with the passion and patience of his grandfather
Never complaining once about snagging lily pads
Casting to the exact spot where the big bass was waiting
Helping to clear the table after supper without being asked
Adding one smooth grey almost-perfect heart shaped rock to
        the cabin collection
Capturing the heart of his grandmother to add to his own
        collection

Sternman

She was but three when she walked straight into the sea on the
    beaches of Weymouth
Salty froth tugging at lacy little socks
Mary Janes filling with sand
Splashes reaching the hem of her Sunday-best yellow pinafore.

Her father preserved the sweet moment in a blurry photo
Her arms extended as if to catch a wave that might fling itself into
    her embrace.

A half century hence she has walked straight into our lives in Down
    East, Maine
Lean and fit and eager as a teenager on a first date
The new lobster traps are pinafore yellow fitted with shinning aqua
    warp
Each waiting to swing into her outstretched arms.

The captain winches them up and places them there for her
And in her new life, it is she who is embraced by the sea.

With Grace

I have never seen you frown
In spite of the deep knowledge
That must bubble and ferment
Behind those mischievous eyes.

Knowledge is one thing
Understanding, quite another
But to create framework
Connecting the wildly disparate, is a miracle.

How can simple buttonholes
Form wormholes to travel through time?
Secrets of ourselves waltz through our eyes?
Freckles become contagious?

You keep me on edge
I don't understand; no-I really do understand
How you must approach the day at a slant
One thumb aloft; beautiful thoughts streaming.

Shadow Puppet Show

Early this morning, Bonnie told me that she had just returned from
skiing on freshly fallen snow
She and her exuberant canine companion
And when she looked up at her lovely cottage silhouetted road-side
in its snowy blanket
Full of all the memories and mementoes of a long and well-lived
life
She had the stark realization that if by some strange quirk
A fire suddenly consumed the entire cottage and all inside
It would be a quite manageable grief and she would just move on

As if to her
The feeling of snow drifting quietly against her face
Her sense of the wild enchanting strength of the natural world
around her
Her fervent embrace of its inherent challenges
All those feelings were what really mattered most

And those feelings are but complex patterns of our brain cells
Nudging each other.

Not reality – they are too fleeting, too ephemeral, too unpredictable
to be considered as such
Rather a type of silhouette, perhaps
Able to capture in our complex neuronal circuits
Something even more valuable than any given object
Creating a shadow of ourselves
Cast upon the rich tapestry of the world around us

Why, He Asks

If I write,
I can seek out essential realities
To capture them
And hold them
As if they were nestled in my hand

If I write,
I can let chaos free my thoughts
To scale their orders of magnitude
From exquisite detail, to utter vastness
With repeating motifs oscillating between strange attractors

If I write,
I can call forth a concentrated calmness
To sooth and order
The brutality dealt to me
By the rawness of which this world is made

If I write,
I can love you in the full complexity
That the years and their excursions bring upon us
I can sort through the tangled swirl of our Wild-To-The-Heart*
bond
And follow the curing lines and letters
Home.

*Rick Bass, *Wild to the Heart*, 1987.

Poems From My Porch

Everything is a poem
Even though the poet, Mary Oliver*
(Whom I think of as a poem herself)
Says, not so.

Ha.

Sweet touches from the night wind under waning stars of winter,
    not a poem?
Silent explosions of a fox mousing on freshly fallen snow?
Silken fragrances from daffodils drenched in spring-melt?
Sibilant eruptions of a wren singing out its nesting boundaries?

Unspoken
Unframed perhaps
Nature's poems nonetheless
Everywhere

Ha!

*Mary Oliver, *New and Selected Poems*, volume two, 2005.

HARD WINTER

My friend, Nancy used to say that she loved weather. Not just certain types of weather, but weather itself; the insistence of it (others would say the inconvenience of it) in our lives – jet streams, the planetary tilt and resulting solstice and season, El Niño and La Niña, squalls, cold fronts, doldrums, dog days, Indian summer, and blackberry winter. Today we had weather Nancy would have loved – hard winter. It was late in coming this year. Flora and fauna have been peaking around the corner of January to see if spring were really here so early. Daffodil buds have been in place, poised, for two weeks. Maple trees sport fuzzy halos of leaf buds. Yesterday, winter answered with a blast of north wind bringing snow and sleet to put spring on hold.

The hillsides and meadows, the run-off stream and ridges appear resigned, patiently bearing the cloak of winter in placid brown and grey. The occasional pine or cedar tree and sprigs of strangle vine offer the rare glimpse of green. Shifting snow showers cover this bareness with a subtle short-lived dusting of white, lost more to sublimation in the wind than to melting. The trees bend and sway, causing bare limbs to scrape and press against each other with low moans and creaks, and smaller branches vibrate with a shrill whistle that rises and falls with the wind. It all seems so somber and serious. Against this backdrop, the little winter birds appear as comic-tragedy – frantically darting, repeating call notes, flitting around trees, canvassing top to bottom desperately for food to fuel their vulnerable metabolism needs. They must ride the whims and wishes of weather, follow its chaotic course through the shift of seasons. They must read the messages in the passage of stars across the heavens, the angle of the sunrise and moon set, and match the rhythms of their sweet small lives to what they can discern. Weather is their lifeline.

That is one of the comforting things about weather. It offers a reminder of our rather insignificant place in the grand scheme of things. It reminds us of our own vulnerabilities. It allows us to pause, to pay homage to the workings of the universe, to seek refuge in our home base, to reaffirm what is important, and to relinquish – albeit momentarily – our compulsion to be *in control.* Well, we are not *in* control. We are simply, *in.* We are participants in the playing out of time and place. According to the poems of Mary Oliver,* this deep knowledge is the principal blessing of life. We do not have to be good or to beg forgiveness for our shortcomings. What is important is that we listen to the call of wild geese flying overhead so that we can comprehend our place in the world – our belongingness. With that anchor, our lives can unfold with grace, with purpose and meaning.

*Mary Oliver, *New and Selected Poems*, volume two, 2005.

Red Oak

It lay massive and still at the edge of the meadow
The only movement coming from a slight rustling of leaves
Caught and crushed among its fallen limbs
A soothing breeze in the soft glow of morning light
Offering balm to the fallen oak

The storm had appeared in darkness
A howling scream as two thunderstorms collided
The force of the explosion broke the storms apart into small,
        powerful downdrafts
Individual violent bursts creating patches of destruction
Like huge footprints left by an angry giant

The red oak was one of many we lost but it was my husband's
        favorite
It had stood for more than two centuries dominating the small ridge
        beyond the lake
Its broad canopy arching above lesser trees
The vanity of its lush adornment of leaves and acorns had caught
        sail-like in the wind
And was its downfall

Our younger son determined that he should tend to the oak by
        himself
He made his way through tangles of broken branches
Climbed along the expanse of bare trunk
Deepening his knowledge of the life of the tree
And its catastrophic death

Days of solitary work sawing and stacking larger limbs for firewood
Placing smaller branches into brush piles to be burned or left for
wildlife cover
The girth of the denuded trunk near its base reached above his
shoulder
The mass of Its upended roots was broader than two measures of his
out-stretched hands
Both trunk and roots remain still, as sylvan sculpture

The other remainder is the acorn mast richly decorating the forest
floor
A feast for deer, squirrels, blue jays
Who in their forays will inevitably distribute acorns to neighboring
ridges
Warmth from oak burning in the fireplace this winter; new oaks
sprouting in the spring
A remembrance and a promise

Monica's Journey

I can see you in my mind's eye
Gliding down over the humps and bumps
Blond hair blowing across your young face
As you glance back at your father
To see his proud, reassuring smile

And then turning to face forward
Your own captain
Your own ride
But to this day
You still take him with you

(And I love being in the wake.)

The Couple

He cares for her
Not that she is needy
But his gentle concern adds a symmetry, a completeness
That sculpts and expands the complex bond between them.

He is ever the gentle man, honoring his father's path
Resolutely devoted to preserving traditions
She still places chains of clover flowers across each path she takes
Still looks through those lovely, mesmerizing eyes as if she can see
to your soul.

He – lying prone on the ancient midden of Volcano Bay
Propped on one elbow, Elmer Fudd cap slightly askew
With characteristic patience, he searches and finds a button
From a Russian soldier, lost from centuries past.

She – sharing confidence and commitment to her bond with
wilderness
As she places her arms around a friend and novice fisherman.
Under her guidance, together – they set the hook
And feel the magic of a wild creature held on a string.

Our lives have intersected; the many crossings celebrated and
revered
We have joined them in their anniversary dance on a wooden deck
near the White River
Shared the haunting of barred owls at Ward Lake, as wood ducks
flew overhead to roost

Floated the Futaleufú River by day and listened to "Musico-
Classico" during magical evenings

Over twenty-five years of adventures large and small
Yet, we cannot recall an argument
Just a remarkable sharing of golden moments
Bonded in rare and lasting friendship.

On the Istanbul Bus

Six small bauble bracelets
Pink, blue
On a slender arm
Reaching to grip the rail for balance

Her upturned gaze
Innocent, confident
Locks on my eyes
Silently extracting from me what she desires to know

Under her youthful stare, I smile
Helpless, exposed
Unable to filter what I am telling her in silence
Trapped in her intense observation

She looks to be shaping her vision of a new world
Cobbled, interwoven
Some of her information coming from the likes of me
As we both contemplate our connected future

Uncovering

Walking the cobblestone streets near Istanbul University
He moves with a loose swagger
In his wake, a covered female figure.
He progresses without a backward glance
Confident of her shadow following.

But then she stumbles,
Emits a small, stifled sound he does not hear
She falls headlong on rough stones.
Folds of black fabric flare
The full extent of her bare legs, pale and plump
Revealed

I do not turn away as modesty would expect
But watch this brief moment of exposure
While the husband turns in obvious annoyance
To attend his wife's recovery
Her shroud descends to re-envelope her
Her form erased

This scene unfolds beneath large, bright posters
Proclaiming recruitment opportunities at the University
Plastered with images of eager and unguarded faces
Of the young women students
Smiling.

Jason's View

A tattered doll – abandoned
Dress torn, scraggly hair in disarray
Why seek this as the telling photograph?

Nothing uplifting, inspiring, joyous
What does it have to say of interest to us?

Perhaps something deep and important
About what we risk in our unpreparedness
About lessons hopefully learned
About courage to face our mistakes

There are strong feelings evoked by the photograph, but of the
darkest shade
Calling to mind the playful young girl
Who might have been gleefully embracing this symbol of her
anticipated role as an adult
At the exact catastrophic moment when that anticipation vanished

Chernobyl.
Healing and resolve are the only answers.

IN LIEU OF VIOLENCE: PREDATION, EMPATHY, LOVE, SPIRITUALITY AND CHOICE

(In tribute to Stephen Pinker)

She seeks
Not blindly or on whim or with predatory malice
But with bits of reasoning nested together
Pointing to things of real value
Fulfilling, affirming, and pleasing

He empathizes
Not with closed mind or limited vision
But with open reach
To those with ties
That fuel imagination in expanding circles

You love
Not limited by passion or drive
But flamed by reasons both subtle and solid
Engulfed in belongingness
To reach beyond who you are

I worship
Not in deference to religious authority or dogma
But in reverence for the nature of things
In wonder of universal workings
That resonate with my soul

We choose
Not by force or random flip of the coin

But with jewels of reason and hard-earned wisdom
Together forming a lighted path to define us
Thinking and therefore being

ENLIGHTENMENT

Awakening
Eyelids slowly lifting
Somnolence fading
The mind's eye view begins to focus on the what and where of the
        day

Things seen and felt provide an incomplete picture
Lacking the phantasmagorical details
Which lie beyond the common senses
Things that are not as they might seem

Beneath the sleep-tossed tresses and in its protective cradle
Lies grey matter, moist and mushy
Pulsing with calculations
Carried out by flowing electrons underlying consciousness

Above the cover of rumpled sheets
Space zooms outward beyond the bedroom walls,
Reaching through the skies where there is no air, only lightness
With occasional galaxies clustering around their black holes.

The mind's view must stretch to embrace the unseeable
Lying beyond the narrow slip of universe we comfortably inhabit
To fashion photons and sub-atomic particles in order to probe the
        origins of our ailments
To harness the power of our star in order to reach more distant ones

We must be wise enough.

Our children’s children’s children
Will need this wisdom to continue to flourish as a species
To seek a stewardship of the universe we know, and that of which we can only dream.

LOVE OF SCIENCE

THE WONDER OF LEARNING

The summer has come and gone. With its end, there is the uneasy expectation of change. Blazing hot days are interspersed now with an occasional cool afternoon shower or a slight briskness in the first air at dawn. Fall is approaching and the smell of it wafts on still, cool mornings at least for a short while until the early September sun burns it away. But those few hours release us from the hot grip of summer, stirring us with as yet faint promises of the autumn wind – sharp, crisp and clean.

As do most things in the South, the weather changes slowly, with a certain grace. No abrupt switch, no rough edges. That would be unseemly in this culture which values the slow, measured pace. We slip from the mesmerizing hold of those stretched-out summer afternoons and the evenings of smothering softness. We give a bow and curtsey to summer's sun-paled skies before we turn our tanned cheeks toward the chilled blue-bright heavens. It's as it should be, a dance of the seasons.

My autumn dance is the dance of academe. Since the age of four, the Fall has marked me for the beginning of a new year of studies. As the summer is politely ushered out on cooling breezes, the boisterous and bright-eyed student in me is called forth to begin anew. I don't know where this inquisitive child hides July through August; it has a secret hiding place in my psyche. Sometimes I mistake it for dead. Then a never-failing clock chimes once more and the child raises its tousled head, stretches out its weakened arms and begins to scan the horizon.

Preparations begin with the gathering up of things, timeless and familiar. Clean white sheets of paper – loose-leafed, spiral-bound and in thick tablets – with their smooth and sleek textures, they await assault from black, fine-tipped pens. Soft pulpy sheets with widely-spaced lines will serve pencils, newly sharpened and with full erasers not yet rounded by a scholar's indecision. Shiny paper clips rattle in their boxes and become tangled in their own repeating design. Brightly colored tabs and dividers will form barricades against the untidy flow and swirl of knowledge. And books, smelling new, with pages clean and uncreased, await their virgin reading. Most prized are the thick text books which boast heavy padded covers and sturdily stitched bindings to withstand the eager fingers that will carefully search the page, flip them impatiently and wear them down with repeated strokes over useful passages.

So armed, the awakened scholar will eye the horizon because this is where the challenge emerges. Fall will appear on the horizon, off in the distance moving slowly and bringing into view the unknown, the out-there, the to-be-known. The attack is launched as always: with a cleared throat, a muffled cough, the slight shuffling of paper, and "Now, class, we begin!"

It matters little who leads the attack. It is an unimportant detail. Students teach professors as well as professors teach students. The quest is paramount. Games and parades of our academic institutions are a rich backdrop for the adventure. But in the end, learning is a solitary achievement: one mind reordering its circuits, making associations, enhancing its ability to predict out of chaos the right answer, or the outcome of any one of life's experiments. The achievement is marked quietly, by a knowing smile slowly spreading across the face or hands stretched out and placed behind the head, as the chair tilts backward. It is a simple moment, it changes your life. You know a new thing; you understand it.

From the countless orderings of individual minds, a collective energy erupts, flooding the campuses of schools everywhere. A great chorus of voices swell, old in tune with the young. It is a mighty force. The sleeping giant awakens from idle summer days and shines in full raiment, bursting forth with a cloak of autumn colors. It is the great enterprise of our species. New insights are breathed into life and spread among all who wish to know of a new thing.

And if you are not too much in a hurry; if you take the time to enjoy the dance; if you close your eyes now and again to rest and reflect, then the rhythm of the seasons will help carry you through many repeating cycles, slowly as the fading of a Southern summer, with grace and as it should be.

STOLEN GLIMPSES AT THE MEETING

When no one else was watching
I saw
A distant rainbow
A broad-tailed hummingbird in mating dance just outside the window
Venus rising

When our young students hoped everyone was watching
I saw
Their bright eyes searching
Their eager hands reaching for a future only they can imagine
Hopes rising

When I was called upon as colleagues watched
I feared they saw
My slowed step
My sentimental recall of things past and hesitating grasp of present
Doubts rising

When others are passing the torch beyond my watchful eyes
I hope to see
A lingering view of those roads traveled
A blue damselfly resting quietly on my hand
Venus always rising

A FIRST DAY

I sense this as the first true day of retirement from my biomedical research career. Perhaps one in a succession of first days. This particular first day celebrates the joy of distraction, of taking tangent paths. The progression of sparkling ruffles I see moving north to south on the surface of the lake easily takes me along for the ride. The journeys continue through the re-reading of diary entries and poems just as effortlessly as the geese glide along the shoreline in a statement of their forever mating-bond.

In my search for strengths to be found upon turning seventy, I am hopeful that the joy of distractions may be among those hoped-for benefits. Along with the phantasmagorical confusion of so many memories, maybe there is an emergent ability to navigate them without the frustration and impatience of earlier years. Single-mindedness has lost some of its appeal. Rather, the messy discordance inherent in a longer lifetime provides a pleasing complexity for those who are prepared to embrace it. This acceptance is freeing to me, as if a gate has swung open to a new playground beckoning.

THE JOY OF QUESTIONS

A really good poet can take the humblest observation
Then contemplate, probe, dig to the core
To reveal mysteries, unknowns
Startling in their revelations

What we don't know often teachers us more than what we do know
Answers will always be incomplete
Questions are our lovely companions
Inspiring us toward pathways of discovery

All you need is to find are the right questions
And (if you are lucky)
That really good poet among your friends
Or the one who resides within you.

GO WITH ME

Could I interest you in cosmology?
Even though its limits are too tiny and too large to fathom
From the tiniest bits of matter – heterologous strings, quarks and
        bosons
To the unfathomable infinity of multiverses?

Would you wish to envision things larger than Mt. Katahdin
Towering in the distance?
Smaller than the single strand of blond hair you brush back
From your granddaughter's brow?

If our thoughts and reasoning can show us
There is much more
That lies beyond our direct observation
Would you like to know?

Would that enrich your life, your dreams?
Would you be pleased by knowledge of your seamless connections
To things larger and smaller, more perplexing
Than you physically encounter?

Would this quest help us find our way in the world?
Would you go with me to that place
– Strangely known and unknown –
And keep me company?

RUNGS OF THE LADDER

Poets dream
Free from bonds
Weaving wisps
Creating baskets full of promise

Theorists formulate
If so, then what?
Following trails left or right
Finding salient crossroads

Experimentalists examine
The impossible implications
Exploring ideas
Matching them with human observation

Progressives incorporate
Integrate and implant
Finding symmetry between what is and what can be
Fulfilling the poets promise

THOUGHT UNIVERSES

And so,
Are we a community
That walks around
With "little bangs" going on inside our heads?

Do we mimic the Big Bang
By creating personal thought universes
Exploding with new ideas
That expand through our endless succession of brain waves?

Do our individual swirls
Of reverberating neuronal firing patterns
Generate new order, new virtual representations
Whose hypothetical total exceeds the number of particles in the universe?

How many cycles of deep theta rhythm brain waves are played out in one's lifetime?
(10 hertz times 60 seconds times 60 minutes times 24 hours times 365 days times 85 years = 26,438,400,000)
How many bits of information are phase-coupled in each theta brain wave broadcast?
Are we humans alone in this activity?

When I occasionally tire of the company of other people, I remind myself of the ongoing miracles in each of their brains. Each and every one. They deserve my respect. In truth, I am in awe of the individual, self-contained power of representational information that resides within every single human being, and as far as we can currently comprehend, ONLY within each of us. Exceptions are likely to be extremely rare.

OUR TIME AND PLACE

We reside in a narrowly confined and unique temporal niche:

Close enough to the beginning of time
To still see the cosmic footprint
Left by the event horizon of the Big Bang
(allowing us to gain important clues as to the nature of our universe)

Far enough away in time
To have had the opportunity
To evolve to our current state of sentience
(Enabling us to contemplate such things)

But not too distant in time
So that matter would have become so dispersed
By the expanding universe, as to be invisible
(Preventing us from seeing any remnants of the surrounding
cosmos.)

We occupy a narrowly confined and rare spatial niche:

Matter packed close enough for a planet
Whose gravity is just right to hold our feet firmly on the ground
Yet permit escape trajectories when fueled by sufficient energy
(allowing us to jump over an obstacle or to the moon as we wish)

Our distance from the sun
Close enough for a suntan and synthesis of necessary vitamin D
But far enough to avoid sizzling

(offering usable levels of energy to fuel our endless ideas)

A silver blue jewel of a planet
Imbued with a lively mix of oxygen, water and complex chemicals
To drive the endless replicative reactions we call life
(Evolving through small changes and new complexities to produce –
us)

At present, I occupy the red couch in our remote cabin in Maine with my husband, listening to "Rhapsody in Blue" on the portable CD player and hoping the fog will lift by midnight so we can see the Perseid meteor shower bright against the darkest skies of almost anywhere else in America. And during such moments that somehow feel propitious, I wonder about the particular circumstances that led me and my fellow humans to occupy this apparently unique position in space-time.

CONVERSATION

We have most certainly arrived at this point
Together as a species.
We hold tight to each other
Addicted to constant chatter among ourselves.

We engage in incessant information transfer
By language, music, images,
Visual cues in body posture or appearance,
Touch: a pat on the back, a shove, a hug, a nudge

But mostly it is by language
That we learn of the world,
Share with others,
Garner plans to shape our environment.

Our insights, theories, jokes, insults,
Shared through language can sculpt our future actions
With such power and with such consequence
That we dictate choices for our planet.

Even now as I have these thoughts, I am using language for a conversation with myself – in my own head. It is, after all, the most comfortable way to think. With words we enter a virtual world to process and manipulate as we please. The evolution of our ability to manage our lives and our environment was surely driven by communal influence through long-lived, multi-generational family units extending to tribes, societies, populations. The logarithmic gain of language-linked brains processing input/output, abstracting key elements, deciding "What next?" must have been an explosive force selecting for the rapid, complex communication of the spoken word among like-minded members of the species.

BRAIN TALK

It must have taken a substantial leap to make a new type of sound
wave
Not just a warning cry, mating call or a mothering coo to trigger
reflexive behavior
Rather, sounds that could recall representational objects in time and
space
Strung together with grammar and logic.

Words directly engage circuitry in the brains of all who would listen.
Through the soft puff of air sent coursing through your throat as
spoken word
Or the facsimile you make permanent through hard copy
You can enter the brains of others, as I do yours at this very moment
Without so much as a touch.

ENTANGLEMENTS

Yes = one bit of information
No = one bit of information
Maybe = something more interesting
A blurred edge
Uncertainty as to where yes ends and no begins

Who cares?
Rocks don't care; it is strictly yes or no with them
Yes = I roll
No = I stay
Maybes don't count

Who cares?
Life cares
Life imbues itself with maybes
Maybe you would like to dance with me
Maybe we could explore our futures together

Maybes count because they imply entanglements
Entanglements imply there are connections between entities
Connections imply we are not alone
But rather, hopelessly enmeshed – hopefully enmeshed
Holding on, celebrating uncertainties

SYMMETRY

Quietly gather your thoughts
Seek a secluded space
Close the door for a while
And leave the world outside

As you sit alone to muse with your inner self
It becomes clear that in fact
The world never leaves you
It remains within you forever, defines you

In your mind's eye
You see and experience
Much of your life's journey
From your knowledge, to your impressions, to your dreams

As if you have captured the things important to you
As a stack of images, reflections
Forming a definition
Of who you are: selected bits and pieces of the world around you

Perhaps not perfect mirror images
Each has been shaped and retouched according to the moment
But these glimpses retain a key element:
Your symmetry with the natural world

At the heart of our existence
(And that of everything else we are aware of in the physical world)
Is a basic symmetric organization

For every elemental particle, there is a mirror image: a matching anti-particle
There is even a hypothesized, yet-to-be-discovered anti-universe to match our very own

Even for intermediate-sized objects (larger than quarks but smaller than universes)
Attributes are not random
Rather they are predictable with a given probability
And thus can be exact symmetric mirror images or (more commonly) replicates of highly similar objects (groups of atoms, elements, planets or human beings)

Two complimentary strands of DNA separate
Then replicate a symmetric structure
Thus preserving the blueprint of life
Symmetry applies to actions as well as particles
"For every action, there is an equal and opposite reaction"
But as conditions change, those actions/reactions can change
Resulting in probability (rather than complete predictability)
Thus affecting your ability to make that hole in one.

Lots of interesting things happen as a reflection of symmetry
The world is not chaos because symmetry provides basic order and predictability
Our minds can help us navigate our world because
We do not have to decipher each object/action as unique

We are able to recognize groups of like objects
So that if we see one representative, we can match it with one of our stored images
And predict its attributes and actions.

Uniqueness is over-rated

The familiar is much more instructive, useful, comforting
Our brain's trillion synaptic connections evolved to seek and encode
repeating objects and actions

Our language is based on the pairing of sounds with groups of
similar objects
Poetry reaches even further
To find deep symmetries between the sounds and feelings of words

We dance to the repeating symmetries of music
We find pleasure in the hidden symmetries of art
We search for a matching soulmate
As a reflection of who we truly are.

LITTLE TIME – BIG TIME: A COSMIC REFRAIN

Perhaps there are two extremes of time:

One, which would mark each spiraling cycle
Along the trajectory, the lifespan
Of an individual oscillating particle
Ticking along like a sub-atomic clock

The other, which would mark the movement
Traversing the infinite lifespan
Of an evolving universe
Swaying back and forth through its seasons.

Somewhere in between these two extremes
I sit, quietly aware of my own repeating heartbeats,
My own rhythmic breathing,
And intimately feel the steady progress of time's arrow.

So it seems
All things small and large
Happen over and over
In their own time

Be they the ten-Hertz oscillatory brain waves
Thought to give us sentience
Or the eternal cycle of creation and collapse
From universe to black hole and back again.

But these are not identical repeats:

Each cycle differs slightly from the last
Each new heart beat, each breath is unique
As if they were evolving
Leaving marks in the fabric of time.

Keeping time
Universal rhythm
Each human heart beat is unique
reflecting relationships
Content there are ways of thinking about the cosmos
That resonate with both logic and feeling, data and intuition
Giving meaning to time, marking time with their unique signatures
We can say they evolve.

Evolution
At atomic, biological and cosmological scale
Marks the passage of time
As a fundamental principle of nature.

RUDDY TURNSTONE

A bewildered little ruddy turnstone has just come by. He stepped lightly on the south Texas beach, halting to quickly poke his bill in the sand, then rush on, head jerking back and forth, to reach the next inspection site. He came toward me in a more-or-less straight line and seemed not too irritated by the human barrier in his path. He then curved around me, as he maintained a distance of ten feet. Trajectories of most things in nature, including that of the ruddy turnstone, are rarely straight. Clouds, wind, waves, the pattern of palm fronds, sand dunes, pelican wings, a young girl's bikini-clad hip, the sun-hastened lines of an old man's face. These trajectories in space are all curved. Are trajectories in time curved as well? The simple laws of entropy say no, but the long view of the space-time continuum hint otherwise. If time is relative maybe, just maybe, it curves a bit occasionally. The lack of curvature, it seems to me, is based on maintaining a constant distance relative to some fixed reference point. If the reference point is moving as well, how can you keep score? All bets are off. Our tacit acceptance of time's straight-forward march is reinforced by our own obsession with marking time in uniform segments. Seconds, hours, work calendars, birthdates, bicentennial reunions, on-time airline departures, everyone in step. We are irritated when we have to accommodate nature's penchant for the ebb and flow, wax and wane aspect of reality. Resetting the nuclear clock at Greenwich, daylight savings time, the unpredictable length of the rainy season or the day of the last frost before spring. These natural cycles, which are not always predictable or entirely constant, offer other hints of the potential curvature in the path of time's arrow. Most of us would in fact agree that time appears to pass at an uneven pace. For me, time passes like a stream rushing white and frothy over rocks at one point in the river bed, then slowly fanning out into deep slow-moving pools below; sometimes reaching into meandering branches that dead-end or curve back to the main channel, creating eddies and back-washes at its juncture. Maybe this perception of time results from reliance on reference points that are themselves not fixed; that are themselves moving, changing. Like my emotional state: joyous, and time rushes past; apprehensive, and time flows into slow moving inlets and dead-end coves that stagnate; sad, and time curves back on itself in self-replicating eddies, making it difficult to break through on a straight course again. Mine is a meandering journey leading to Sky Top, Tennessee, and now, to a visit on this south Texas beach, with reference points emotionally tied to my past experiences, to who I am and their molding of who I am, to my relationship to my family, to my husband – all of whom have their own curving paths. The energy required to remain on a straight and narrow path is enormous and it is energy unwisely spent. The sharp boundaries created by unyielding, unresponsive resolve are counterproductive to natural progress. You will get there, we will get there, I will get there just the same if I respond to the detours and the meanderings along the way, follow the circling of hawks that soar in the summertime, stop to check the scrapes along deer trails, reach with my fly rod into the quiet backwash along the lake shore, sit quietly to watch the sun fatten in parallax on the western edge of the horizon, study the movement of my husband's hands as he carefully dresses our freshly killed deer in the light of the lantern I hold in the approaching evening darkness. This is how time moves. I have let go – and let the passage of time carry me along its curving paths.

TIME AND AGAIN

Two words powerful enough to explain the universe.

The present moment in time reflects all there is.
Complex conjectures of probable connections of events
Linking to the past – or pointing to the future, thus create the present (right now) And all there is.

Probabilities are real; not entirely certain but sufficiently suggestive
Probabilities of the sun rising; the seasons turning; our growing old
For convenience, we imagine an arrow of time to provide
A unidirectional roadmap of existence driven by probabilities

And again, if present events are driven by high probabilities
Aren't those events most likely to happen again?
Suggesting that life itself is a series of high probability repeats?
The molecules of life directing a symphony
A repeating rhythm of cell divisions, growth, death
And then, starting anew

Life emerged from a backdrop of monumental repeating events, big and small
Planets continually twirling on their axes,
Electrons orbiting atomic nuclei
The rhythmic repeat of vibrating strings possibly creating the fabric of all existence

I gleefully skip rope to the repeating heartbeat of the universe
Time and Again

TIME IN A BOTTLE

The glass bottle, tightly capped, encloses a once gloriously-colored
maple leaf
Now dark brown and enshrouded by a beard of grey and white
fungus.

In the *past,* this wizened sprig was first spring green, then autumn
red
Decorating branches that arched over a path winding through the
wild Maine woods.
One prized leaf gathered and shared.

These events, in their trajectories through Space-Time,
Leave memory traces, mental maps in my brain
To enrich my enjoyment and remind me of their meaning.

At *present,* the bottle sits in its special spot on my desk
Triggering interactions with my memory traces
These interactions occupy (create?) this particular instance of the
present.

In the fabric of Space-Time, Warp – Woof defines Where-When
If positions are close enough (in space) and simultaneous enough (in
time)
Events can occur to create the present point in Space-Time

Light reflecting from the bottle is absorbed by my eye and makes me
smile at 2:54p, Wednesday, January 22, 2020,
creating the present moment.

And for the *future,* the forward trajectories of events,
There are no memory maps
Only probabilities for the changing configurations of Space-Time
and the elements in it.

Probabilities are shaped by universal entropy
Relentlessly moving order toward disorder
The incessant ticking of the cosmic clock.

Will we spend another summer in Maine?
Will the entropy of advancing age dissipate our desire
For encounters with loons, crashing waves on the Bold Coast, the
splash of Milky Way seen against the pure night skies?

THE MEASURE OF THINGS

The Heisenberg Uncertainty Principle invokes the most human of traits, measuring, in an explanation of the predictability of matter. In the grand scheme of the multiverse, does our small compulsion to gain a measure of all things actually doom us to ultimate failure because our instruments are too insensitive, clocks too slow to obtain measurements of a given object in space-time simultaneously? The Uncertainty Principle says no, rather it is the nature of space-time itself that precludes predictions of these absolute values with absolute certainty at an absolutely determined time.

Even if reaching the level of absolute predictability on a quantum scale is not allowed by the laws of physics, still there remains an infinity of interesting measurements to be made. Normally, we miss out on a lot of these because of scale. Many of the really intriguing things are very tiny and very fast. These small, quick elements of our universe are doing curious things – appearing out of nothing then returning to nothing in an instant; occupying all possible states as a sum of its histories; participating in the infinite expansion of our universe. We have gained this understanding of what is otherwise hidden by observing the behavior of by-standers. For example, the matter we can see behaves as if there is much more out there that we can't see (dark matter).

Pondering these possibilities can make our billions of synapses hum. Billions are still not infinity, but maybe close enough for practical purposes. The sum total of possible combinations of firing patterns of those billions of synapses is larger than the number of particles in our universe – so, close enough. We may be virtually unlimited in our conceptual, sensory and motor memory. And, perhaps not surprising, there may be a bit of cosmology in the workings of our brain: In mimicry of our expanding universe, our memory clearly expands during our lifetime; collective memory also likely expands during the lifetime of our species as a result of evolutionary selection. Like particles and anti-particles, ideas (meaningful synaptic firing patterns) can be created from nothing (background noise) and just as easily return to nothing – as if they never existed. Sums of histories as a general concept is also familiar to us since we continually play the game of considering all possible choices, creating a virtual sum of histories, before deciding on a plan of action. As members of the human species, we have an additional attribute that makes us different from other conglomerates of elemental particles: we encode measurements of things around us and use some of the cosmological principles just mentioned to create mental representations or virtual images that are manipulated in mental space-time. We use these manipulated measurements to plan our day, to decide what to do next. This act of deciding is thought to imbue us with sentience. Consciousness.

So, if this is how it works, I question our current lack of integration of neuroscience into cosmological measuring principles. If we are using these billion synapses of ours to gain a measure of all things – even the most minute and ephemeral – should we not keep in mind *The Mind* that is doing the measuring? We know in some detail the common neural pathways, how they are wired and how they operate on a molecular level; we know of some of the computational limitations of neuronal networks (i.e., 300 msec are needed to make a novel observation); we are gaining an understanding of the types of hard-wired thinking our entire species participates in, not all of which are useful or even germane (i.e., evolutionary relics). Surely this neuronal substrate influences our measuring. By nature, our views have to be parochial – unless we factor in (or rather factor out) the influence of the "measurer" – us – from the measurements and resulting theories.

Yes, I think we are clever enough to do this. It would be refreshing and instructive to know our universe from, say the anti-particle point of view. Jumping fully formed into existence, an absolute mirror of a not-yet-met perfect partner particle somewhere out there, and in the blink of its sub-atomic eye encountering the ultimate, consummating collision to enter nothingness again. See how easy it is to anthropomorphize? Imagine how challenging it would be not to.

FIRST MEMORY

It would be easy for organisms to determine what to do next if in fact the environment never changed. "Just keep doing what is it you are doing" would be a good strategy. But life on a planet that twirls on its axis and orbits its sun is exposed to continual, cyclic change: hours of exposure to the sun's energy followed by hours of darkness, with the periodicity modified by the seasons. To gain full advantage of the free energy upon which all life on earth depends, it would be advantageous for an organism to predict when the next change in the cycle will occur. Based on that prediction, cells could gear up or gear down accordingly and get metabolic machinery in the proper mode to absorb energy or to assume a rest mode.

The fact that virtually all living earth creatures do exactly this type of prediction, suggests that its benefits drove selection early in evolution. The prediction mechanism is known as circadian rhythm. Clock genes operate as a delayed time loop, triggered/entrained by light but once initiated can free run in the absence of light for a while until damped out.

I have thought that clock genes were such a good idea, worked so well for their intended purpose, that nature may have used a similar blueprint for other important activities. Like memory. In one sense, memory serves as a device that guides us in deciding what to do next, and gets us ready for the change, just as circadian cycles of melatonin and dopamine continually get us ready for photopic change. Perhaps the memory trace is similar to the workings of clock genes: sensitized by some internal or external input, triggering metabolic/ionic changes *with a time-delay* that keeps us poised for making the next step – not *permanently* poised – but in preparation mode for appropriate periods of time before damping out.

My thought is that circadian rhythms may have served as the primordial memory trace, maybe light and darkness formed our first memory.

TREASURE THE MOMENT

Not sure what *time* is. In all probability, we humans invented *time* as
a way to navigate (thanks, Mary Soon Lee*) the only
universe that exists: the *present.*

We are able to roughly calculate/describe events in the universe as
bits/elemental particles that are either present (+; in the
present) or absent (–; not in the present).
So perhaps, the *present* is all there is; a succession of *presents.*

Each successive step (consisting of particles deciding to be or not to
be) is influenced by the tapestry of the preceding *present.*
Voilà: *probability.*

Snapshots of the *present* can reflect the past as well as portend the
future (via the probability tether)
Creating the impression of a moving image of time
Just like in the movies.

As a further complication
Each elemental particle or group of particles can be influenced
differentially
Depending on its own individual properties.

Whether your train is moving toward or away from the sound of the
station whistle,
Will determine the pitch of the sound you hear.
(Isn't that what *relativity* is? – but that's another topic.)

By developing designations for *before, during* and *after*
Our human brains may have created the notion of a time-line strategy
Allowing us to understand the past – accommodate to the present –
predict the future
All based on *probability.*

Probability is real. *Time* may not be.
What are the chances?

*Mary Soon Lee, *How to Navigate Our Universe*, 2023.

RETROSPECTIVE/PROSPECTIVE

What can the past tell us about the future?
Neither past nor future exist
Rather, existence is uniquely the feature of the present.

The past fades away; the future forms possibilities; we remain
eternally poised in the present.

Past and future are but probability maps
Tied like strings to the passing moments of the present.

Ah, but what fascinating strings, in mesmerizing patterns that enrich,
challenge, amaze

All would be beyond our grasp
(Given our caged existence in the present)
Save for the ability of our own internal neuronal circuits
To interpret and comprehend the advancement of our entangled
existence
Through the rushing waves of time.

ANATOMICAL HERESY

Picture if you will taking a high-resolution image of an orchestra
playing Mozart,
using that image captured in one moment in time to map the position
of each finger of each flutist, each violinist, each musician
then using that map as a basis to discover music theory.

In my estimation, we put ourselves in roughly the same position
when we try to discover a theory of mind
based on a snapshot-derived, anatomical map of where synapses are
observed
at any one point in time.

Neurons have a special beauty of their own,
free-form cell bodies sprouting thin tentacles to reach amazing
distances
touching other neurons much as a musician's fingers might press
violin strings
constantly changing their synaptic touches to create the music of our
minds.

Without incorporating the fourth dimension, time
our attempts to develop theories of thought
remain frozen within a single, small, and fixed instant
ignorant of the phantasmagorical melodies of the cerebral symphony.

SINGULAR

One smooth grey pebble tumbles against countless others on an
endless beach
A single frothy wave marches among legions of white-capped
colleagues to reach the far shore.
One among many, but still uniquely one; one separate bit
Upon which the cosmos and its complexities depend.

I am also a oneness, a unitary being
Tumbling and marching through time with fellow singularities
Creating the vast historical entanglements
Written on our universe.
.
One is more than just a number
Rather a type of deep, defining quality
That can embrace a distinct history and evolutionary path
That can forge a distinct future of burning dreams and bizarre,
unexpected connections.

You can mathematically determine the number of miles to the moon
Tally them as 238,857 in preparation for the journey
Yet grasp a truer understanding of the distance as a single unit
"One small step; one giant leap for mankind"

Some argue that the universe itself is singular;
Not multiple, not infinite, not timeless
But one continuous fabric woven together by each single reality
Advancing through a secession of moments.

I accept oneness as our shatteringly beautiful gift from the natural
universe.
Each of us uniquely constituted and embedded in time
Never to be exactly duplicated or repeated
One surrounded by many; entangled, intertwined, and thus defined.

Four Percent is Plenty

We know so little, we are told
Our universe holds much more than we are aware of – maybe even
more universes
Apparently (at least to physicists) huge stores of matter and energy
surround us
Hidden in darkness.

And so I sit
Peering through the thus-limited human peephole
At what appears to be pervading bounty
Oblivious to what might be filling all the "empty" spaces

Most pleasing shapes of puffy white clouds
Suspended in the endless blue sky.
I remain unaware of the dark matter
Which could obscure these visual treats.

The joyful calls of red-breasted nuthatches, belted kingfishers.
Might they be drowned out
Awash in the swirl of dark energy
I cannot sense?

Do we really need to observe the full measure of all there is?
The apparent absence of some things, the spaces thus created
These seem important to me, not just dark nothingness
Rather a guide for distinguishing what is truly here and now

As we face the loss of a dear friend

The lack of his presence in our lives
Brings with it a true realization of all that he has meant to us
A deep perspective gained only from shadows and echoes of the past
splayed open
Allowing a healing acceptance of treasure no longer there

EVERYTHING FROM NOTHING

Nil
No thing
No thing ness
Empty ness
Total darkness
Absolute zero; absolute stillness
Silence

Out of the unstable void, enter
One
Quantum
Bit
Wave particle

Add/subtract
Divide/multiply
Integrate/calculate
Follow the never-ending numerals of pi

At Einstein's speed of light
Logarithmically expand to include Heisenberg uncertainties
Create Reinmann math for the Standard Model
Discover Hilbert space vectors for elemental vibrating strings
Examine black body radiations and black hole holograms

Celebrate
N
Fin
I
T
Whee!

IN FORMATION

Entities that assume a form
Non-random groups
Creating a formulation of energy and matter plus space

These patterned particles of energy-matter can be translocated
But why bother
Easier to transfer the pattern rather than the particle

Transfer by transforming the pattern onto other particles
In particular, other fast-moving wave particles like electrons and
photons
The patterns can then leap through space

Following predetermined paths
Seeking appropriate targets
Thus your sister's telephone number appears on your cell phone
screen

MORE?

You may say, "But we are more than a string of numbers"
Math is lovely, powerful yet you perhaps disagree with
Max Tegmark* when he suggests that
"The universe is nothing more than a mathematical construct."

However, the theorem states there are truths
Which lie beyond any given set of axioms
Thus leading to evolution of new theorems.
An infinity of evolving theorems

Perhaps that is all there is in our mathematical universe
If I am some small part of that process – Wow!
If that is all I am and there is nothing more – Fantastic!
I would not ask for more.

*Max Tegmark, *Life 3.0*, 2017.

KNOWING

An idea born of human contrivance
Plants don't know
Animals don't know
Humans don't know either

Rather we have used our massive 10^{15} bit circuitry
To calculate probabilities
For mass, energy
And changes in entropy

We call these probabilities "knowing"
But they are only rules of thumb
That can predict, sometimes with exquisite precision
The exact moment the sun rises every morning

WHEN? + WHERE? = WHY?

Trajectories define all of existence
From elemental particles to solar systems to universes.
The flow of all objects follows a path defined by perceived
movement
From previous position (past)
To current position (present)
To projected position (future)

Only two trajectory-related data points are possible:
Where the observation was made
When the observation was made.
A series of these paired measurements,
Define a probabilistic trajectory for the object.

Trajectories are of interest only to living creatures
(Rocks don't care.)
The difference between rocks and living creatures
Is based on the unique ability of cells (specifically their DNA)
To self-replicate, and in doing so, to evolve.

Cellular networks, particularly those in the brain,
Calculate and compare trajectories;
Then adjust behavior to maximize success, ensuring that genes
related to that success
Are selectively passed to future evolving generations.

These calculations have also achieved something remarkable,
Forming the basis of what is possibly a uniquely human feat:

Namely the ability to pose the question, “Why?”

Other animals alter their behavior based on direct observation and
reflex response;
But there is no evidence they do so based on patterns of “When?”
and “Where?”
That underlie the search for broad-reaching constructs for
understanding the universe.

“Why?” is perhaps uniquely human.
I wonder why that is?

SELF

The replicative unit
Endowed with the ability to reproduce itself in its entirety
Toes, eyeballs, kidneys
A somewhat loose organization of bits and pieces, each with
different design plans

Differing sets of genes are expressed differentially, including all
genes required for replication of the individual cell or the
fused product of two replicative cells, male and female
To do the job, our species requires only about 20,000 active, protein-
encoding genes
All internally or externally orchestrated

Each gene is a selected survivor of truly countless contests
You represent the cache of current winners
That allows you to succeed in your niche

Our actions are coordinated to act out our genes' instructions
In order to meet "our needs," namely, "our gene's needs"
Paramount of which is to prepare for successful replication of the
gene set

We do not have Free Will.
Our brains calculate probabilities of proposed actions
And we always choose the action most likely to successfully pass on
our genes
It feels like Free Will because the various probabilities are so
complex

Superficially, our motivations do not appear to be linked
To a defined set of reproductive strategies
But they are predictively linked, at least to some varying degree of probability
The rest is noise

Is "self" your physical being at present?
Or is it the set of all human genes and all their possibilities
Contained within the full collection of human inhabitants?

Certainly, the two are linked,
Namely, what is good for your physical being
Can often be predicted by what is happening to fellow humans

Likewise, what is good for the set of human genes as a whole
May also be determined by analysis of interactions of individual humans.
The trend of human evolution appears to support
This merger of individual versus species gene selection and resulting behavior selection

Call it morality? We have moved toward a more global view for determinants of our actions
Be it control of climate change or decline of tribalism and violence

Will our global view extend to other planets, to other lifeforms?
How will our selfish genes respond to aliens?

WHO KNOWS

Why am I doing this?
Not a question I ask often, but occasionally
Say, when I have faced a steep incline up a mountain pass
Seeking solitude that suddenly smothers me

How, what, where, when
Pose easier questions
Offering definitive answers
(General Relativity and Heisenberg's Uncertainty Principle,
notwithstanding)

But why
Begs deeper, confounding responses
Answering: "because of…."
Only begs the next query: "Then, why that?"

In one sense, never-ending
In another sense, the acceptance of the question is its answer
The peak of the mountain gained; the view embraced as a given
The quest for why is who we are

PERSONAL RELATIVITY

Awakening to a new day
The mass of grey matter inside your skull
Sets its circuits to fresh incoming news
Each millisecond's worth of data, new and old, to be reconciled

This thing that is you
Is but a calculation
As tenuous, as heartbreakingly naive, as breathtakingly magnificent
As any aggregate of matter ever imagined

You find your way
Through your sentience
Gathering the ones and zeroes of the quantum cosmos
Into symmetries that constitute personal experience

Your own time and place represent the only absolute
A specific Space-Time in which you cause events to occur
All else is relative, existing in separate constructs of their own past-
present-futures
No two entities experience things identically

A lonely perspective perhaps
A realization that your views might be mistaken
Yet empowering; endowing profound uniqueness
Marking your presence in the ever-changing cycle of being.

Brains

Bat-sized brains tuned to echoes,
The capacious memory of an elephant matriarch –
Able to map water holes for her troop,
Curious dolphins conversant in clicks,
The vision of hawks sufficient to discern moles a mile distant

Instances of mental magic
Honed by eons of evolution
Elegantly linked to the external world.

We have some of this, too,
In bits and snippets and traces:
A connecting cord, a brain stem, a diencephalon
But dwarfed and encapsulated by additional brain stuff
Uniquely human
At once beautiful and burdensome

Our oversized neocortical layers
Umbrella-sized if spread out
But tightly scrunched in the cranial vault
Large enough
Complex enough
To encode the world and bring it inside

Virtual data points assembled
Mentally manipulated
Filtered by learning, culture, language, art
Skewed by incestuous introspection

How perplexing the connections
Between old real-world brain and new thinking brain
The interface of paleo – and neo –
Seeking poetic reconciliation

A HARD MISTRESS

In truth, you only have about sixty seconds; less than a full page
That's it. That's all you get
To share your heart and soul; your statement of the world; your true
confessions
It feels like trying to write a *poem*-mercial

Like you are reaching to briefly tap someone on the shoulder
Saying, "Hey, have you ever thought of this?"
No more Odysseys
Just quick thrills; then time to move on

And the rules are hard, like a maze
Hidden deep, diverse and layered
But we all know they are there, when they are stretched
And when they are broken

Why has poetry evolved as the major remembered voice?
In song, or dance or chant or recitation?
Now I lay me down to sleep
With this ring I thee wed
Blue, blue, blue suede shoes, baby

It is something that strikes to the soul.
"I recognize you – you are a writer. A poet?" she said to me
And squeezed my heart
The heart of my hard mistress and me

ACKNOWLEDGEMENTS

Original bird photographs by the author and copies of corresponding haikus were purchased by customers at the Eastport Gallery, Eastport Maine, 2019.

Reprint permission has been granted for these previously published writings:

"The Sky" appeared in the annual anthology of high school poetry by the National High School Poetry Association, 1957.

"Wesley Pine" appeared in the *Machias Valley News Observer* in 2009.

"Sweet Barter," "Holding On," "Uncovering," and "Belonging" appeared in *From a Far Corner: An Anthology of Poetry from the Easternmost Reaches of Maine*. The Primavera Press, 2018.

About the Author

My family background includes generations of red-neck farmers on the northwest Louisiana/Texas border (on my father's side), plus Cajun market hunters from Bayou Blue outside New Orleans (on my mother's side), all of whom gave me freedom and encouragement to seek my own path, which turned out to be dominated by a fascination with scientific searches for the right questions and right experimental techniques to expand our understanding of ourselves and the universe we inhabit. *Love Notes* reflects that continuing journey. Retired after a total of 50+ years as a vision research scientist, initially at the University of Texas Health Science Center in Houston, Texas, and subsequently at the University of Tennessee Health Science in Memphis, Tennessee, I am now spending my eighth decade in small rural communities (one in Tennessee during the winters; one in Maine during the summers) that offer abundant opportunities to celebrate nature, time spent with loved ones, and writing.

www.ingramcontent.com/pod-product-compliance
Lightning Source LLC
LaVergne TN
LVHW010725061126
851857LV00014B/1347